European Research Activities in Cloud Computing

European Research Activities in Cloud Computing

Edited by

Dana Petcu and José Luis Vázquez-Poletti

CAMBRIDGE
SCHOLARS
PUBLISHING

European Research Activities in Cloud Computing,
Edited by Dana Petcu and José Luis Vázquez-Poletti

This book first published 2012

Cambridge Scholars Publishing

12 Back Chapman Street, Newcastle upon Tyne, NE6 2XX, UK

British Library Cataloguing in Publication Data
A catalogue record for this book is available from the British Library

ISBN (10): 1-4438-3507-2, ISBN (13): 978-1-4438-3507-7

TABLE OF CONTENTS

Part II: Cloud Management

Part III: Cloud Automation

Part IV: Cloud Adoption

CONTRIBUTING AUTHORS

Imad M. Abbadi, University of Oxford, UK
Antonin Abhervé, SOFTEAM, France
Mohammed Airaj, Linear Accelerator Laboratory, Orsay, France & Caddi Ayyad University, Morocco
Owen Appleton, Emergence Tech Ltd, UK
Franck Barbier, University of Pau, France
Marc-Elian Bégin, SixSq Sàrl, Switzerland
Arne J. Berre, SINTEF, Norway
Alysson Bessani, University of Lisbon, Portugal
Per Brand, SICS, Sweden
Antonio Bucchiarone, Foundation Bruno Kessler, Italy
Sven Bugiel, Technical University of Darmstadt, Germany
Cinzia Cappiello, Politecnico of Milano, Italy
Roberto G. Cascella, INRIA, France
Emanuele Cesena, Politecnico of Torino, Italy
Ilknur Chulani, ATOS, Turkey
Maria Couceiro, INESC-ID/IST, Portugal
Francesco D'Andria, ATOS, Spain
Michel Dao, Orange Labs, France
Mina Deng, Philips Research Europe, Netherlands
Elisabetta Di Nitto, Politecnico of Milano, Italy
Karim Djemame, University of Leeds, UK,
Albert Eckert, Siemens AG, Corporate Technology, Germany
Daniel Espling, Umeå University, Sweden
Evangelos Floros, Greek Research & Technology Network, Greece
Georgina Gallizo,High Performance Computing Center Stuttgart,Germany
Spyridon V. Gogouvitis, National Technical University of Athens, Greece
Sergei Gorlatch, University of Münster, Germany
Michael Gröne, Sirrix AG, Germany
Danny Harnik, IBM Haifa Research Lab, Israel
Piyush Harsh, INRIA, France
Matti Heikkurinen, Emergence Tech Ltd, UK
Christian Hein, Fraunhofer FOKUS, Germany
Alexis Henry, BLU AGE Software and Netfective Technology, France

Michael C. Jaeger, Siemens AG, Corporate Technology, Germany
Yvon Jégou, INRIA, France
Miguel Jimenez, Politécnica University of Madrid, Spain
Gregory Katsaros,High Performance Computing Center Stuttgart,Germany
Stuart Kenny, Trinity College Dublin, Dublin, Ireland
Dimosthenis P. Kyriazis, National Technical University of Athens, Greece
Elliot K. Kolodner, IBM Haifa Research Lab, Israel
Kleopatra Konstanteli, National Technical University of Athens, Greece
Ignacio M. Llorente, Complutense University of Madrid, Spain
Charles Loomis, Linear Accelerator Laboratory, France
Mirko Lorenz, Deutsche Welle, Germany
Ninja Marnau, Unabhängiges Landeszentrum für Datenschutz , Schleswig-Holstein, Germany
Dominique Meiländer, University of Münster, Germany
Andreas Menychtas, National Technical University of Athens, Greece
Alberto Messina, RAI - Centre for Research and Technological Innovation, Italy
Andreas Metzger, University of Duisburg-Essen, Germany
Parastoo Mohagheghi, SINTEF and Norwegian University of Science and Technology, Norway
Rafael Moreno-Vozmediano, Complutense University of Madrid, Spain
Brice Morin, SINTEF, Norway
Christine Morin, INRIA, France
Srijith Nair, British Telecom, UK
Dalit Naor, IBM Haifa Research Lab, Israel
Stefan Nürnberger, Technical University of Darmstadt, Germany
David O'Callaghan, Trinity College Dublin, Dublin, Ireland
Eduardo Oliveros, Telefonica Investigation e Disarollo, Madrid, Spain
Sebastian Ortega, University Politécnica of Madrid, Spain
Marcelo Pasin, University of Lisbon, Portugal
Dana Petcu, West University of Timisoara& Institute e-Austria, Romania
Muttukrishnan Rajarajan, City University London, UK
Tom Ritter, Fraunhofer FOKUS, Germany
Luís Rodrigues, INESC-ID/IST, Lisbon, Portugal
Paolo Romano, INESC-ID/IST, Lisbon, Portugal
Andrey Sadovykh, SOFTEAM, France
Thor Sæther, DI Systemer AS, Norway
Ali Sajjad, British Telecom, UK
Norbert Schirmer, Sirrix AG, Germany
Aidan Shribman, SAP Research Israel, Raanana, Israel
Alexandra Shulman-Peleg, IBM Haifa Research Lab, Israel

Michał Śmiałek, Warsaw University of Technology, Poland
Javier Soriano, Politécnica University of Madrid, Spain
Philipp Strube, CloudControl, Potsdam, Germany
Sivan Tal, IBM Haifa Research Lab, Israel
Johan Tordsson, Umeå University, Sweden,
Jose Luis Vazquez-Poletti, Complutense University of Madrid, Spain
Salvatore Venticinque, Second University of Naples, Italy
Athanasios S.Voulodimos,National Technical University of Athens,Greece
Oliver Wäldrich, Fraunhofer SCAI, Germany
Yaron Wolfsthal, IBM Haifa Research Lab, Israel
Wolfgang Ziegler, Fraunhofer SCAI, Sankt Augustin, Germany

PREFACE

Cloud computing is currently undergoing a great deal of discussion. However, despite considerable efforts and investments in the industry of cloud computing, critical problems have yet to be solved and the research and development community involved in distributed computing is searching for viable solutions.

This book intends to present a snapshot of the on-going activities in research and development undertaken at the European level through collaborative work of multi-national teams in the frame of the FP7 programme of European Commission.

The material is split in four parts. Two are oriented towards the potential clients of cloud computing: on one side the service offer towards the potential clients of cloud computing, on another the methods of adopting the cloud computing paradigm. The other two parts address the cloud computing providers' problems: one deals with the cloud management while the second with the automation efforts.

The idea to group in a book the contributions of different European projects dealing with cloud computing problems has emerged on the occasion of organising the second Workshop on Software Services: cloud computing and Applications based on Software Services (2nd WoSS, June 6–9, 2011, in Timisoara, Romania, http://sprers.eu/events/2nd-woss).

In a period which is confused about what cloud computing can and cannot achieve, we hope that the content of the book will transform those clouds in the reader's sky into white, fluffy and friendly ones.

We warmly recommend this book to the students, researchers and developers working in the field of distributed computing.

—Dana Petcu and José Luis Vázquez-Poletti

INVITATION TO A JOURNEY IN THE ERA OF CLOUD COMPUTING

DANA PETCU

Before starting the adventure of solving the current cloud computing problems together with the authors, the reader is invited to a briefing on cloud computing and its status in European Research Area (ERA).

First Step is Also the First Problem: What is Cloud Computing?

A common understanding of the notion of cloud computing is still missing. This fact reflects a natural process in computer science, and especially in distributed computing. A similar example is that of grid computing, a term coined two decades ago and an acceptable definition of which has come about only in the middle of the last decade.

Taking into account the orientation of this book towards exposing the research and development activities in cloud computing in European collaborative projects, we are considering here the European Expert Group definition (from its report The Future of Cloud Computing. Opportunities for European Cloud Computing beyond 2010, January 2010):

> A cloud is an elastic execution environment of resources involving multiple stakeholders and providing a metered service at multiple granularities for a specified level of quality (of service).

This definition underlines two of the main characteristics of the cloud computing paradigm which differentiate it from other distributed computing paradigms: elasticity and metering.

The definition promoted by NIST (US National Institute of Standards and Technology) from January 2011 reveals another characteristic—on-demand provisioning:

> Cloud computing is a model for enabling ubiquitous, convenient, on-demand network access to a shared pool of configurable computing resources (e.g., networks, servers, storage, applications, and services) that can be rapidly provisioned and released with minimal management effort or service provider interaction.

The cloud computing paradigm shakes up the market of IT services by its pay-per-use concept for the leased resources and its promise to cut operational and capital costs. The resources, either infrastructure or software, are available over the network and their providers use a multi-tenant model and virtualisation techniques.

The deployment models of the cloud computing concept are currently classified as follows:

- Private Cloud, when the resources are owned or leased by a single organisation
- Community Cloud, when the resources are shared by several organisations between which agreements are established in topics like mission or security policies
- Public Cloud, when the resources are sold to a general public
- Hybrid Cloud, when a composition of several clouds of previous types is built for a service or application with the aim of providing a new business, enable data and application portability, or support elasticity when needed.

The resources in cloud computing are delivered as services. The "As a Service" (with the well known acronym aaS) paradigm refers to components across network which are available for reuse. A large variety of resources can be offered following this paradigm, leading to a general image of "anything" available "as a Service" (shortly XaaS). Hardware, software, platform, infrastructure, process, storage, database, identity, application, integration, network, monitoring, testing, high performance computing, communication, governance, security, backup, marketplace, and even human experience are nowadays offered "aaS." However, the most common service delivery models are the following three:

(1) **Software as a Service** (SaaS), also known as Application as a Service (AaaS), is the model in which an application is hosted as a service and customers access it using Internet protocols. It is the opposite of the Software as a Product model, currently used worldwide. Already, classical examples of SaaSs have emerged,

including office automation applications, accounting systems, online video conferencing, web analytics or web content management tools.

(2) **Platform as a Service** (PaaS), also known as Cloudware, offers the resources required to build new applications and services, without locally installing software. The current offer includes services for application design, development, testing, deployment, hosting for team collaboration, services for database integration, scalability, and versioning for state management.

(3) **Infrastructure as a Service** (IaaS), also known as Hardware as a Service (HaaS), allows the renting of CPU cycles, server space, network equipment, memory and storage space. The major benefit is the fact that the infrastructure can be dynamically scaled up and down depending on the users' needs.

Cloud computing technologies

While cloud computing is a new way of delivering resources, several old and new technologies and tools support the paradigm. The high interest in it, as well in its financial benefits has resulted in a large variety of technological offers. Any attempt for their classification will be absolute at the moment of its appearance, since the market is still highly dynamic.

With regards to this introductory material, we propose a simple classification based on several criteria of a refined delivery model, like hosted service or deployable services, distributed or single site offers.

Most well known in the cloud computing market are the hosted services, available to all delivery levels, SaaS, PaaS or IaaS, that are accessible as Public Cloud offers. On the other hand, for the developers of new services and applications, the deployable cloud-native or compliant services and tools are important assets. Since this book reports on the research activities of European computer scientists involved in the daily innovation of new technologies and tools for cloud computing, deployable services are a major topic.

Tables I.1 and I.2 below propose a taxonomy of the existing offers of cloud computing services, split at the highest level in offers for resources (Table I.1) and offers for services and application development (Table I.2). These tables provide just a snapshot of the available offers without trying to include all of them or to offer a straight split between categories (since several technological offer are spread over several categories). They intend only to help the reader of this book to identify the type of technologies that will be referred to in the following chapters.

Table I.1. A partial taxonomy of the resource offers available for cloud computing in the summer of 2011

	Offer type	System	Group	Representatives
Storage	Columnar databases	Hosted		Amazon Simple DB, BigTable
		Deplo-yable	Distri-buted	Cassandra, Hadoop HBase, Hypertable, Tiramole
			Single server	M BD, MonetDB, LucidDB, Akiban
	Key-value databases	Hosted		Amazon S3, RackSpace Cloud Files
		Deplo-yable	Distri-buted	Riak, Membase, kumofs, LightCloud, GridFS, Voldemort, Redis, Hibari, Kal, Ringo, Dynomite, SubCloud
			Single server	MemcacheDB, Tokyo Cabinet, Kyoto Cabinet, BerkleyDB, LevelDB, GT.M
	Document databases	Deplo-yable		MongoDB, CounchDB
	Distributed file-systems	Deplo-yable		Hadoop HDFS, Tahoe LAFS, Ceph, Sector, CloudStore, MooseFS, mofilefs, XtreemFS
	Distributed atomic databases			Zookeeper, Scalaris, Keyspace
	Distributed hash tables			Kademlia, Pastry, Tapestry, Chord
Compute	Map-Reduce	Hosted		Amazon Elastic Map Reduce
		Deplo-yable		Hadoop, Riak Map Reduce, Hive, Hadoop Pig, Cascading, CloudBase, Sawzall, Disco, Http Map Reduce, DyradLINQ, fairly
	Infrastruc-ture	Hosted		Amazon EC2, Amazon EBS, RackSpace Cloud Servers, GoGrid, Force.com, FlexiScale, Slicehost, ElasticHosts, CloudSigma, NewServers
		Deplo-yable		Eucalyptus, OpenNebula, OpenStack, Nimbus, XCAT, Cobbler, BoxGrinder, Tashi, VPN Cubed, Elastic Server, OpenQRM
Communicate	Message queues	Hosted		Amazon SQS, Amazon SNS, StormMQ, Linxter
		Deplo-yable	Clus-tered	RabbittMQ, ZeroMQ, HornetQ
			Single server	ActiveMQ, MorbidQ, Kestrel, MemcacheQ, Starling

Table I.2. A partial taxonomy of the developemnt offers available for cloud computing in the summer of 2011

Support	Type	Representatives
Platforms	Hosted	AppEngine, SmartPlatform, Heroku, Azure, GigaSpaces XAP, Stackato, CloudIQ, Cloud Burst, Jamcracker, Bungee Connect, LongJump, Enomaly ECP
	Deployable	AppScale, TyphoonAE, Cast, Project Caroline, SpawnGrid, Duostack, CloudFoundry, NodeJitsu, Nodester, DotCloud, CloudBees, OpenShift, Aneka, AppZero, vCloud, RightScale, JOALA Cloud Manager
APIs	Libraries and frameworks	DeltaCloud, DataNucleus, Spring Data, libcloud, CloudLoop, Dasein Cloud API, jclouds, Simple Cloud API, Restlet, Sun Cloud API, BOOM, SalsaHpc, Dryad, Orleans
	Standards	OCCI, CDMI
IDL		XPIDL, ETCH, WebIDL, Ice

Table I.3. Communication protocols

Support	Type	Representatives
Message queues	ASCII	STOMP
	Binary	AMQP, JMS
Serialization	Binary	Protocol buffers, Erlang external term format, MessagePack, Avro, BSON, XDR, ASN.1
	ASCII with binary support	UBF, S expressions, Bencode
	ASCII	JSON, XML, Netstrings, YAML, SDL
RPC	Binary	Sun RPC, BERT, Hessian, Thrift, AMF, RxPC
	ASCII	JSON RPC, XML RPC, SOAP
	Queues, deployable	Geaman, peafowl, beanstalkd

Tables I.1-3 indicates the communication protocols that are currently used in cloud computing.

What in Missing in the Cloud Computing World? Towards a Research Agenda

The main challenges for cloud computing identified in the Open Cloud Manifesto 2009 were data and application interoperability and portability, governance and management, metering and monitoring, as well as security. Note the projects that will be presented in this book emerged in the context of that year's concerns: the research proposals have responded to the call for projects elaborated under the umbrella of the moment-related needs of cloud computing that were specified above. Therefore, the reader will find in this book several ideas that take a concrete form in dealing with the complex conceptual and technical issues behind the main challenges of cloud computing.

The identification of new challenges is a constant topic both for researchers and practitioners. In this context, Armbrust et al. (2010) revised the top ten obstacles and opportunities for cloud computing. The primary obstacles identified are: data lock-in, confidentiality, auditability and transfer bottlenecks; quick scaling, performance unpredictability, scalable storage, management of large distributed systems, reputation sharing, software licensing and business continuity. Despite the concentrated efforts of the cloud technology developers, most of these problems still remain unsolved and are subjects of research agendas. In this context, the top ten opportunities mentioned by Armburst et al. are the following: use of multiple cloud providers, compatible software to enable hybrid clouds, special encryption and firewalls, higher bandwidth switches, improved support for virtualized resources (including debuggers), scalable store, build auto-scalers, snapshots acquiring techniques, reputation-guarding services, and pay-per-use licenses. In particular, the subject of using multiple clouds has revealed considerable interest, also reflected in this book.

Concerning the recommendations for the European Research Area in cloud computing, the European Expert Group in cloud computing stated in 2010:

> Europe's main opportunities to participate in the cloud movement consist in particular in aspects related to extending and completing the capabilities of current cloud systems, whereby the long-term goal consists in realizing meta-scalable cloud systems and services.

ERA of Cloud Computing

A new era in distributed computing was opened with the advent of cloud computing. Worldwide computer scientists, researchers, programmers, IT specialists, analysts, teachers, CEOs and even politicians are using the new paradigm in their daily activities or communications with the hope of contributing to some solutions of our daily societies' problems, through innovation or commercialization, or support of both.

The tremendous activities in cloud computing are recognised by experts designing the European Commission's programmes for research and development. Moreover, their analyses show that the era of cloud computing will be at least as long as that of the grid. Therefore, the Seventh Framework Programme (FP7) of the European Commission stimulates research, development and innovation in the cloud computing area through indication of the high expectations on: research and development activities in the context of the Cooperation programme in the Information and Communication Technologies (FP7-ICT), transfer of knowledge, and innovation or support activities in Ideas, People or Capacities programmes. Other European Commission's programmes, like Competitiveness and Innovation Framework Programme (CIP) or Cooperation in Science and Technology (COST) are also looking forward to applications related to cloud computing.

The landscape of current European Research Area (ERA) related to cloud computing is diverse, varying from new emerging concepts and technologies to simple usage in innovative applications. In what follows we will focus on the activities reported in the frame of FP7, organised in the context of projects of different sizes, from integrated projects (IP) and networks of excellences (NoE), to small targeted research projects (STREP) and support actions (SA). Fig. I.2 below presents in this context a snapshot of the clusters of the FP7 projects related to cloud computing. Note that this snapshot reflects the state of the art as of summer 2011 and it is developed with the occasion of preparing this chapter (not to reflect the opinion of the European Commission).

The biggest technological cluster (to which this book mainly refers, and marked as the biggest cloud in fig. I.1) is that developed under the SSAI Objective of FP7-ICT programme (Objective 1.2 from 2009: Service and Software Architectures, Infrastructures and Engineering), where at least ten projects are currently developing the concepts, technologies and tools of tomorrow's cloud computing. Their activities are relying upon the achievements of a previous SSAI cluster (Objective 1.2 from 2007: Internet of Services, Software and Virtualisation).

Fig. I.1. The clusters of FP7 projects related to cloud computing topics as of summer 2011

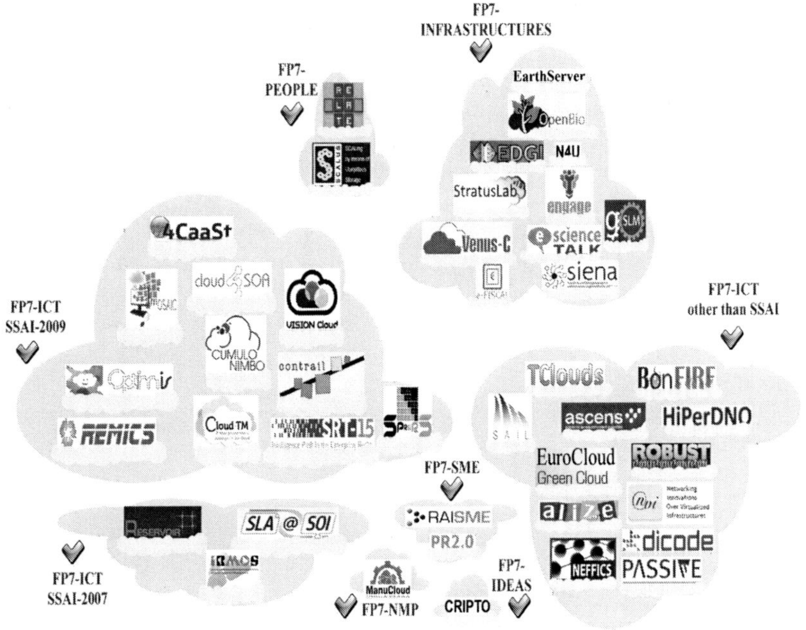

Other objectives of the FP7-ICT programmes are also currently supporting the cloud computing technologies through their enhancement or usage. Their activities are recognised in fig. I.1 as a particular group of twelve non-SSAI projects (i.e. FP7-ICT rather than SSAI).

An important contribution in terms of both technologies, applications and e-infrastructure support is provided by the consortium members of the projects supported by the Capacity programme, in particular the work-programme of FP7-Infrastructure. The specific group of projects is presented in fig. I.1 as the third biggest cluster (FP7-Infrastructures).

Several cloud computing related projects are encountered in other FP7 programmes, like Ideas or Peoples.

The promises and current contributions of these clusters of projects, researchers and practitioners are summarised in the tables I.4 to I.8 below. More details about eleven projects from the three biggest clusters presented in fig. I.1 can be found in this book.

Table I.4. The SSAI-2009 cluster of FP7 projects related to cloud computing topics

Acronym	Web page	Cloud-related topics
4CaaST	4caast.morfeo-project.org	The project aims to create a PaaS which supports the optimized and elastic hosting of Internet-scale multi-tier applications.
Cloud4SOA	www.cloud 4soa.eu	The project focuses on resolving the semantic interoperability issues of cloud infrastructures and on introducing a user-centric approach for cloud compliant applications. It combines three paradigms: cloud computing, Service Oriented Architectures (SOA) and lightweight semantics.
Cloud-TM	www.cloudtm. eu	The project aims at defining a programming paradigm and developing a self-optimizing distributed transactional memory middleware that spares programmers the burden of coding for distribution, persistence and fault-tolerance. The platform features autonomic resource provisioning and pervasive self-tuning schemes.
CONTRAIL	www.contrail-project.eu	The goal of the project is to design, implement, evaluate and promote an open source system in which resources that belong to different operators are integrated into a single homogeneous federated cloud that users can access seamlessly.
Cumulo Nimbo	www.cumulo nimbo.eu	The project promise to deliver a PaaS that will provide consistency, availability, and simpler programming abstractions, such as transactions.
mOSAIC	www.mosaic-project.eu	The project builds an open source PaaS that negotiates cloud services as requested by their users. Moreover, it proposes a vendor-agnostic API for programming and deploying cloud-complaint applications

OPTIMIS	www.optimis-project.eu	The high-level objective is to enable an open and dependable cloud service ecosystem that delivers IT services that are adaptable, reliable, auditable and sustainable. This will allow organisations to automatically and seamlessly externalize services and applications to trustworthy and auditable cloud providers.
REMICS	www.remics.eu	The main objective is to specify, develop and evaluate a tool-supported model driven methodology for migrating legacy applications to interoperable service cloud platforms.
SPRERS	www.sprers.eu	The project supports dissemination of FP7-ICT project achievements related to software services, including organisation of trainings, workshops and this book.
SRT-15	www.srt-15.eu	The project intends to bridge the gap between cloud infrastructures and enterprise services by building a distributed service platform. It relies on four key enabling technologies: content-based routing, complex event processing, dependability and data privacy
VISION Cloud	www.vision cloud.eu	The goal of this project is to introduce a powerful ICT infrastructure for reliable and effective delivery of data-intensive storage services, facilitating the convergence of ICT, media and telecommunications. This infrastructure will support the setup and deployment of data and storage services on demand, at competitive costs, across disparate administrative domains, while providing QoS and security guarantees.

Table I.5. Former SSAI-2007 cluster of FP7 projects related to cloud computing topics

Acronym	Web page	Cloud-related topics
RESERVOIR	www.reservoir-fp7.eu	By merging virtualisation, grid and business technologies, it allowed the migration of resources across geographies and administrative domains, maximising resource exploitation, and minimising costs.
IRMOS	www.irmosproject.eu	Designed, developed and validated cloud solutions allowing the adoption of interactive real-time applications, and especially multimedia applications, enabling their rich set of attributes (from time-constrained operation to dynamic service control and adaptation) and their efficient integration into cloud infrastructures.
SLA@SOI	sla-at-soi.eu	It has provided a framework for enhancing the provisioning of services with: predictability and dependability; transparent SLA management; automation of the process of negotiating SLAs and provisioning, delivery and monitoring of services

Table I.6. The cluster of FP7-ICT non-SSAI projects related to Cloud computing topics

Acronym	Web page	Cloud-related topics
Ascens	www.ascens-ist.eu	The project focuses on service-component ensembles, hierarchical ensembles built from service components, and knowledge units connected via a dynamic infrastructure. To realise ensembles of service components a set of case studies from robotics, cloud computing and e-Vehicles are used.
Aliz-e	www.aliz-e.org	The project develops the theory and practice behind embodied cognitive robots capable of maintaining believable any-depth affective interactions with a young user. The integration of cognitive components is based on cloud computing for embedded systems.
BonFIRE	www.bonfire-project.eu	The project designs, builds and operates a multi-site cloud facility to support applications, services and systems research targeting the Internet of Services community within the Future Internet.
Dicode	www.dicode-project.eu	The project facilitates and augments collaboration and decision making in data-intensive and cognitively-complex settings. It exploits and builds on several concepts like cloud computing, MapReduce, Hadoop, Mahout, and column databases, to search, analyze and aggregate data existing in diverse, extremely large, and rapidly evolving sources
EuroCloud	www.eurocloud server.com	The project focuses on improvement in system density and energy efficiency for data center applications to a scale supporting hundreds cores in a single server. Mobile cloud services are used in validation scenarios, while in general the targeted market is that of the companies able later on to construct efficient, environmentally clean and compact data centers for the deployment of green cloud services.

HiPerDNO	dea.brunel.ac.uk/ hiperdno/	The aim of the project is to develop a new generation of distribution network management systems that exploit near to real-time HPC solutions with inherent security and intelligent communications for smart distribution network operation and management. Cloud and grid computing are used to enable scalable data mining, feature extraction, and near to real-time state estimation.
NEFFICS	neffics.eu	The project builds an innovation-driven ecology for networked enterprises, on top of an established cloud-based, software-as-a-service business operation platform, combined with an advanced innovation management software platform.
NOVI	www.fp7-novi.eu	The project concentrates on efficient approaches to compose virtualised e-infrastructures towards a holistic Future Internet cloud service. Resources belonging to various levels are managed by separate providers. The project search methods, information systems and algorithms enable users with composite isolated slices, baskets of resources and services provided by federated infrastructures.
PASSIVE	ict-passive.eu	The project proposes an improved model of security for virtualised systems to ensure that separation of concerns can be achieved in large scale deployments, threats from co-hosted operating systems are detected and dealt with, and public trust in application providers is maintained in a dynamic hosting environment. It proposes a policy-based security architecture allowing the easy specification of security provisions; fully virtualised resource access, with fine-grained control over device access; and a dynamic system for authentication.

ROBUST	www.robust-project.eu	The project deals with large-scale data management and analysis tasks for understanding and managing complex user behaviours and ecosystems in online business communities. A highly scalable cloud and stream-based data management infrastructure serves to handle the real time analysis of large volumes of data.
SAIL	www.sail-project.eu	The project researches and develops novel networking technologies. It improves application support via an information-centric paradigm and develops mechanisms and protocols to realise the benefits of a Network of Information. It enables the co-existence of legacy and new networks via virtualisation of resources and self-management, fully integrating networking with cloud computing to produce cloud Networking.
TClouds	www.tclouds-project.eu	The project targets cloud computing security and minimisation of the widespread concerns about the security of personal data by putting its focus on privacy protection in cross-border infrastructures and on ensuring resilience against failures and attacks. TClouds develops a cloud infrastructure that can deliver computing and storage that achieves a high level of security, privacy, resilience, and yet is cost-efficient, simple and scalable.

Table I.7. The FP7-Infrastructures' cluster of projects related to Cloud computing topics

Acronym	Web page	Cloud-related topics
EarthServer	www. earthserver.eu	The project aims at open access and ad-hoc analytics on Earth science data, based on the geo service standards Web Coverage Service and Web Coverage Processing Service. It develops open-source client and server technologies scalable to Exabyte volumes, based on distributed processing, supercomputing, and cloud virtualisation.
EDGI	edgi-project.eu	The project aims to deploy desktop grid and cloud services for European Grid Initiative's user communities that require extremely large multi-national e-infrastructure. Software components of ARC, gLite, Unicore, BOINC, OpenNebula, Eucalyptus etc. are integrated into SG-DG-Cloud platforms for service provision. The project also provides a workflow-oriented science gateway to enable user communities to access the infrastructure more easily.
Engage	www.engage-project.eu	The main goal of the project is the deployment and use of an advanced service infrastructure, incorporating distributed and diverse public sector information resources as well as data curation, semantic annotation and visualisation tools. The project explores synergies with e-Infrastructure projects and service providers to leverage on existing grid and cloud services for curating, storing, processing and exploiting large amounts of public sector information.
EuBrazil OpenBio	www.eubrazilopenbio.eu	The project deploys an e-Infrastructure of open access resources (data, tools, services), to make significant strides towards supporting the needs and requirements of the biodiversity scientific community. This data e-Infrastructure results from the federation and integration of individual existing data, cloud, and grid EU and Brazilian infrastructures and resources across the biodiversity and taxonomy domain, namely Catalogue of Life, OpenModeller, D4Science-II and Venus-C.
e-ScienceTalk	www.e-sciencetalk. org	The project aims to bring the success stories of Europe's e-infrastructure to a wider audience. It coordinates the dissemination outputs of European e-Infrastructure projects, ensuring their results and influence are reported in print and online.

e-Fiscal	www.efiscal.eu	The scope of the e-FISCAL proposal is to analyse the e-infrastructure costs, facilitated by national entities (NGIs and HPC centres), compare them with equivalent commercial leased or on-demand offerings and provide an evaluation report.
gSLM	www.gslm.eu	The project aims to improve service level management (SLM) in the grid. It generates new approaches for SLM and offers best practices from e-infrastructure service provision in systems that span multiple administrative domains. The outputs are relevant for managing the QoS aspects of the "inter-cloud" solutions.
N4U	Not available	The project will provide an e-Science environment to deliver a virtual laboratory offering neuroscientists access to a wide range of datasets and algorithm pipelines, access to computational resources, services, and support. It will integrate and augment the service provision with computational resources from grid, cloud and HPC.
Siena	www.siena initiative.eu	SIENA defines a future e-Infrastructures roadmap focusing on interoperability and standards. The roadmap presents scenarios, identifies trends, investigates the innovation and impact sparked by cloud and grid computing, and delivers insight into how standards and the policy framework is shaping future developments and deployments.
Stratus Lab	stratuslab.eu	The StratusLab Toolkit integrates cloud and virtualisation technologies and services within grid sites and enriches existing computing infrastructures with IaaS provisioning paradigms.
Venus-C	www.venus-c.eu	VENUS-C proposes an open and generic API at platform level for scientific applications, striving towards interoperable services. Its platform will be based on Windows Azure, Eucalyptus, OpenNebula, EMOTIVE.

Table I.8. Other programmes' clusters of projects related to Cloud computing topics

Progr.	Acronym	Web page	Cloud-related topics
FP7-Ideas	Cripto	lib.bioinfo.pl/projects/view/29666	The project investigates topics in cryptography from a theoretical and practical perspective. The application domains investigated are cloud computing, electronic voting, protocols for trusted computing and privacy preserving methodologies.
FP7-NMP	Manu Cloud	www.manucloud-project.eu	The project develops and evaluates an IT infrastructure which supports on-demand manufacturing scenarios. It provides users with the ability to utilise the manufacturing capabilities of configurable, virtualised production networks, based on cloud-enabled, federated factories, supported by a set of SaaS applications.
FP7-People	RELATE	www.relate-itn.eu	The initial training network provides opportunities for young researchers to study the latest technologies, platforms and tools in engineering and provisioning of service-based cloud applications.
	SCALUS	www.scalus. eu	The initial training network aims at elevating education, research, and development inside areas of cluster, grid, and cloud storage.
FP7-SME	RAISME	www.raisme. eu	The project builds an open source platform for SME applications. It enables high-tech organisations with niche skills to rapidly build and scale innovative ICT applications. This is achieved through the use of mashup technology and cloud computing.
	PR2.0	Not available	The project promises an affordable, cost-effective centralised web tool for management of marketing campaigns using popular social media tools. It allows companies to monitor user-generated content about their business or brands, to manage their online reputation, and to be alerted to possible public relations crises. A cloud computing model ensure that the tool is accessible to SMEs

Cloud Computing in ERA—Whereto?

The FP7-ICT's Work-programme for 2011–2012 recognises that: "cloud computing is transforming the software and the service industry and can have a profound impact on business ICT strategies in all sectors." Therefore, one of its objectives (ICT-2011.1.2) specifically targets cloud computing technologies, while other objectives are supporting the spread of trust (ICT-2011.1.4) or applications (ICT-2011.5.5). In this context, the main topics of the next project proposals responding to the European Commission call for research and development activities will focus on:

- Hybrid clouds, in particular on topics like interoperability, portability, data protection and its distribution control, service composition across heterogeneous environments
- Intelligent techniques like automatic management of resources, agile elastic scalability, support for context-aware applications
- Improvement of current services and technologies such as those for infrastructure virtualisation, mobile applications, energy efficiency, network support for cloud computing, or open source implementations.

The CIP's ICT-PSP programme for 2011–2012 is also supporting the technological transfer of the results obtained in the frame of research activities towards general society, for example by sustaining the proposals for innovating and making government services more effective and fully interoperable. The emphasis is placed on cloud computing in public services allowing for efficiency gains, service aggregation, sharing and reuse.

On the other hand, long-term research activities in cloud computing are foreseen and recommended by several European organisations of experts, like ENISA (European Network and Information Security Agency) or e-IRG (e-Infrastructure Reflection Group). ENISA has already recommended (in its report cloud computing—Benefits, risks and recommendations for information security, 2009) focusing on building trust in clouds (with a special emphasis on data confidentiality and higher assurance in clouds), data protection in large scale cross-organisational systems (in special monitoring, traceability, forensics and evidence gathering, international regulations of data protection), or large scale computer system engineering (resource isolation mechanisms, interoperability between clouds, improved resilience).

Last words before the journey

As stated in the Preface, this book intends to present a snapshot of the current activities in ERA related to cloud computing, with a particular emphasis on

- nine projects from the FP7-ICT SSAI cluster (4CaaST, Cloud4SOA, Cloud-TM, CONTRAIL, mOSAIC, OPTIMIS, REMICS, VISION Cloud and S-Cube)
- one project from the non-SSAI cluster (TClouds)
- two projects from the Infrastructure cluster (StratusLab and gSLM).

Representatives of these projects met each other to discuss their project status and to identify communalities, overlaps and complementarities in the frame of a scientific workshop (2nd Workshop on Software Services) organised by the support action project from FP7-SSAI cluster, named SPRERS on June 2011 in Timisoara, Romania. Their project presentations in video and written format are available at http://sprers.eu/events/2nd-woss. More details about the projects and their latest achievements are revealed in this book.

The first part of the book is dedicated to cloud services, pointing towards their elasticity and the issues hiding behind the complexity of the technology ensuring the availability of these services. The first chapter presents the vision of Contrail consortium on these topics in the context of federating clouds. Another approach dealing with elasticity of cloud services is described in the second chapter that proposes a new way to build cloud compliant applications (Cloud-TM approach). The third chapter, proposed by VISION Cloud Consortium, focus on data intensive applications and introduces new concepts related to storage in the cloud. The fourth chapter presents the position of REMICS in treating the problem of porting legacy application towards clouds.

The second part of the book addresses issues in ensuring the functionality of clouds. The fifth chapter, presented by OPTIMIS Consortium, dicusses the requirements of efficient cloud services on the data center site. The sixth chapter describes the the cloud approach to building secure, reliable and trustworthy services on top of multiple clouds.

The third part of the book deals with new approaches for building clouds. The seventh chapter presents the mOSAIC's approach for brokering of cloud services and their provisioning at application deployment stage, using agent technologies. The eighth chapter brings in discussion on the potential of self-adaptation in clouds (S-Cube vision).

The ninth chapter (proposed by gSLM) extrapolates, in the context of clouds, the service level management strategies used in the context of other e-infrastructures.

The fourth part of the book is focused on the cloud offer towards scientific and bussines communities. The tenth chapter presents the approach undertaken by StratusLab to connect grids and clouds for the scientific communities. The eleventh chapter provided by Cloud4SOA connects the concepts of service-oriented architectures, clouds and semantics. The last chapter focuses on 4CaaST vision on clouds as business ecosystems.

Summary

In this introductory chapter we have briefly presented the basic notions of cloud computing as the latest paradigm in distributed computing, as well as the context of the projects' activities reported, motivated and sustained in the coming chapters. Each of the following chapters intends to introduce the reader to some of the cloud computing problems for which new solutions are proposed.

We hope that the reader will extrapolate a general image of the highly innovative research European activities in cloud computing or to find a potential answer to his or her questions about cloud computing.

References

Armbrust et al. "A view of Cloud Computing." In *Communication of the ACM* 53/4, 2010.

European Expert Group in Cloud Computing. "The Future of Cloud Computing. Opportunities for European Cloud Computing beyond 2010." Available at http://cordis.europa.eu/fp7/ict/ssai/docs/cloud-report-final.pdf (accessed September 15, 2011)

NIST. "Cloud Computing Programme," 2011. Available at http://www.nist.gov/itl/cloud/ (accessed September 15, 2011)

PART I:

CLOUD SERVICES

CHAPTER ONE

OPEN COMPUTING INFRASTRUCTURES FOR ELASTIC SERVICES: CONTRAIL CLOUD FEDERATION

PIYUSH HARSH, YVON JÉGOU, ROBERTO G. CASCELLA AND CHRISTINE MORIN

Cloud computing is rapidly defining the computing paradigm for the modern networked age. Users can run their large computations online using cloud services at a fraction of the cost compared to setting up their own data centers. Clearly, cloud computing offers many advantages, and yet many large organisations including governments, the financial sector, and the health care sector are reluctant in transitioning to cloud computing.

The contrail project addresses the major concerns behind this reluctance, namely mistrust in cloud platforms, the lack of Service Level Agreements (SLAs) and Quality of Protection (QoP) of data. Contrail aims at providing a federation layer support for bringing a multitude of cloud providers, both private and public, together. This will allow multi-tenancy and cloud-bursting capability to end user cloud applications while supporting SLAs and QoP agreements desired by several privacy aware sectors including governments, banks, health care providers to name a few.

This chapter describes the overall architecture of the Contrail framework; from the cloud federation services to the Contrail Virtual Execution Platform (VEP) that will be closely interfaced with the IaaS layer of cloud providers. VEP upgrades the supported cloud providers and brings trust in cloud computing by adding SLAs and QoP features missing at the typical IaaS layer. The federation services interact with the provider VEPs to provide transparent access to federation resources. Furthermore, this chapter outlines challenges faced in being part of a large federation and how VEP addresses some of those.

1. Introduction

Cloud offerings are maturing with increasingly sophisticated services available to the consumers. End users and corporations have tools available today that enable them to configure and operate their own private cloud services. These tools are available as commercial products, for example VMWare's vCloud (http://www.vmware.com/ products/vcloud/ overview.html), as well as open source IaaS alternatives such as OpenNebula (http://www.opennebula.org/) and OpenStack (http://www. openstack.org/). In addition, there are public clouds that operate on pay-as-you-go model such as Amazon's EC2 (http://aws.amazon.com/ec2/), Microsoft's Azure (http://www.microsoft.com/windowsazure/), and Google AppEngine (http://code.google.com/appengine) into which consumers can launch their applications and services as needed.

The above scenario is definitely a huge improvement from traditional data center models of last decade. Even though clouds offer much better price/performance ratio over non-virtualised environments, the true power of cloud computing can only be achieved with seamless federation among different cloud technologies.

A true cloud federation will bring both private and public clouds under the umbrella of one federation where users will have the option to deploy services using resources of multiple providers. The federation will present to the user uniform API/billing/monitoring features regardless of the nature of actual cloud service providers. Moreover, any organisation can become part of this federation being both a cloud provider, when its IT infrastructure is not used at its maximal capacity and a cloud customer in periods of peak activity. Resources that belong to different operators can be integrated into a single homogeneous federated cloud, shown in fig 1.1, which users can access seamlessly.

According to Windows Azure (http://www.azurepilot.com/page/ Cloud+Application+Types+%26+Patterns), the bulk of cloud applications and services can be categorized into four main types:

(1) **Creating a scalable Web application**. These applications need automatic scale-out and scale-down capabilities in order to adapt the allocated resources to their load. The pay-per-use model of cloud providers is likely to be cheaper than maintaining lots of mostly unused machines.

(2) **Creating a parallel processing application**. Scientific parallel processing applications can demand a large number of computational resources in order to execute in a realistic period of time. However,

sporadic execution does not justify maintaining permanent large data centers and the pay-per-use model can provide a cost-effective solution.

(3) **Creating an application with background processing**. Many web-accessible software solutions need to initiate work that runs in the background, independently from the user-interaction part of the application, for instance for data-mining.

(4) **Using cloud storage from an on-premises or hosted application**. Storing large databases is likely to be cheaper, more scalable, and more reliable on the cloud than on local hardware.

Mixing various types of applications can lead to the optimisation of resource usage at the provider level. According to the scheduling strategies as defined by Haizea in Sotomayor et al. (2008), parallel processing applications can be scheduled using advance reservations, and background applications can be scheduled using best-effort mode with possible pre-emptible leases and/or deadline guarantees. Background processing applications can be pre-empted during peak load to create room for scalable and elastic applications.

The Contrail project (www.contrail-project.eu) aims to achieve these goals by developing an integrated approach to virtualisation, offering Infrastructure as a Service (IaaS), services for federating IaaS clouds, and Platform as a Service (PaaS) on top of federated clouds. Contrail project aims to support, at the federation level, all types of applications that are supported by traditional clouds. This calls for the deployment of a transparent, trusted, and reliable "Contrail federation" with operations governed by Service Level Agreements (SLAs) that provide support for strong Quality of Protection (QoP) and authentication.

No cloud implementation, either commercial or open source, currently implements all features that Contrail will supported: federation, IaaS as well as PaaS, security, quality of service and protection enforced by service level agreements. Open source cloud platforms such as OpenStack (www.openstack.org), Nimbus (www.nimbusproject.org) or OpenNebula (www.opennebula.org/) only support the IaaS layer with no quality of service and protection guarantees. However, some of these platforms such as OpenNebula partially support the notion of a cloud federation.

Full PaaS support is provided by commercial products such as VMware vCloudTM (www.vmware.com/products/vcloud/overview.html) through the OVF distributed application open format (www.dmtf.org/standards/ovf). However, these platforms are currently limited to private cloud offerings only.

A Contrail federation integrates both private and public clouds under a common umbrella. User identities, data, and resources will be interoperable within the federation, thanks to common supports for authentication and authorization, and to a common policy definition, monitoring, and enforcing mechanisms (SLA, QoP, etc.), as well as to a common economic model (Contrail consortium 2011b).

The way the Contrail federation is conceived, together with its open source deployment nature, will enable seamless access to provider resources, avoiding potential vendor lock-in for the end users. These users can now deploy their distributed applications on demand to different providers by negotiating resources with the federation, which will choose the best providers based on the application requirements.

In such a scenario, Contrail technology will be able to satisfy the user needs for the deployment of elastic and scalable applications guaranteeing performance dependability. The run time performance of the application can scale with the number of assigned resources, and the user only needs to specify the desired performance level and Contrail's automatic provisioning system can add resources on demand during the execution of the application to guarantee the desired performance (thus providing elasticity).

A challenging aspect of Contrail is the management of computing resources to guarantee this performance dependability and QoS of active virtual execution platforms. Cloud providers might have different efficiency policies given that the individual resources being contributed to this federated cloud will be highly heterogeneous in their hardware configuration and system-level organisation, thus making their interoperability challenging.

The support for efficient cooperation and resource sharing within cloud federation is critical for the Contrail technology. This is achieved through Contrail Virtual Execution Platform (VEP), an open source technology implementing standards that exploits resource virtualisation to provide distributed infrastructures for deployment of end-user applications independent of the underlying platform. VEP operates above the IaaS layer as shown in fig. 1.1. In Contrail, each cloud provider will run a copy of VEP software that in turn will seamlessly integrate its resources with the Contrail federation.

VEP offers a reliable application deployment platform that is resilient to operational failures and which supports secure management of user data with a strong guarantee for QoS. These requirements are critical in making the provider resources trustworthy so that users can get performance guarantees for their application, and bring trust in cloud computing solutions making them suitable to run user businesses safely.

GAFS, the Global Autonomous File System, will manage Contrail cloud storage (Contrail consortium 2011d). For each user application, Contrail deploys a Virtual Infrastructure Network (VIN), a component that is responsible for managing all communication within a Contrail application. The application appliances and the GAFS file system communicate in a secured way through the application VIN instance.

Fig. 1.1. A simple cloud federation

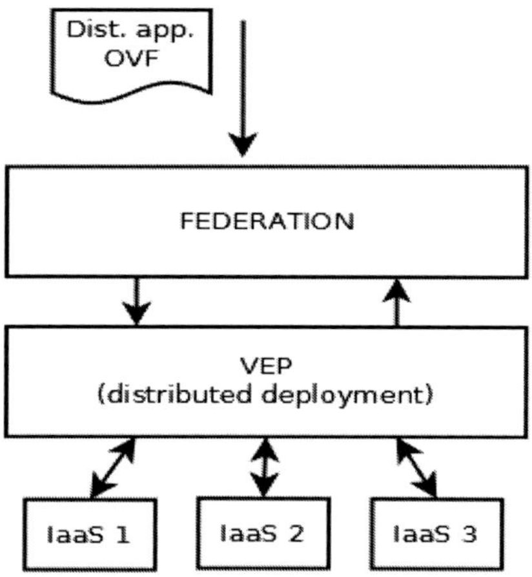

The remainder of this chapter is organised as follows—section 2 outlines the key challenges inherent in a large cloud federation. section 3 describes the Contrail modules and users' interaction with the federation. section 4.1 describes the role of VEP in cloud computing. section 4.2 describes the components and the design rationale of the Contrail VEP development effort. Finally, section 5 describes the Contrail progress report, VEP development road map and recap of important contributions along with the key points presented in this paper.

2. Challenges of a Large Federation

In a federation, many independent members operate together governed by the agreed upon federation laws. Just as in a society where a federation is successful only if the individual members operate properly, the same set of challenges and opportunities exist in a federated cloud. In this section, we will list few major issues that may arise in a federation of clouds. Later on, we will see how VEP helps address some of these concerns.

2.1 Authenticity of Reported Statistics

An end user of a federation of clouds may not know a priori which provider her application will run on. The user agrees to pay the resources used by her application and in turn may require some form of validation of the reported resource usage. The provider may fudge the reported statistics in order to fleece the end users. It would be the task of the federation to ensure such a case does not arise. A clear code of conduct and penalties for violation must be defined, but even then, the enforcement could be problematic. Surprise audits of provider logs by federation officials could be one deterrent against fraud by providers. However, the logs themselves may be modified. Resource providers, federation officials and users can also agree on third party auditors that are then allowed to inspect the integrity of the data. Fool-proof implementation of anti-fraud measures may be impossible to achieve. Further, for a successful federation, the terms of participation for the individual provider must not be too stifling and must not favor large providers over small. Finding the correct participation terms and conditions balance may prove to be key for the success of the federation, and yet may be the most difficult balance to achieve.

2.2 Verification of SLA Adherence

Whenever there is a contract agreed between two parties, the contract enforcement becomes a legally binding obligation. Service Level Agreements (SLAs) in the context of cloud computing could contain agreements on the operating environment and placement restrictions of virtual machines (VMs) at the provider's location. An example SLA object could condition that some VMs in an application must not be placed on the same cluster rack as others. There might be CPU load conditions or memory requirements to name only two such SLA parameters that could be present.

In a large federation where negotiated SLAs are supported between the user and the provider, the verification of the SLA contract adherence could prove to be challenging. If the user deploys their own VMs as part of the application, the user could embed their own verification code to periodically check if the execution environment is within the negotiated

SLA parameters or not. The provider must maintain adequate logs for auditing if the need arises.

The Contrail project will have support for interactive monitoring of applications by the users and administrators in addition to automated monitoring of negotiated SLAs by the SLA monitoring module. There are two major types of SLAs in Contrail, one negotiated between the federation and the individual provider, and the other that a user negotiates with the federation layer for the execution of their cloud application. A Constrained Execution Environment (CEE) in Contrail is the application execution environment by the provider to run user's applications in accordance with the negotiated SLA. Through the federation portal, users can discover the parameters of the CEE derived from the negotiated SLAs on which their appliances have been deployed—on which provider resources, on which physical resources etc. Of course, a provider may deny access to some of this data, but accessibility to this can be made part of a negotiated SLA.

Typically, initial SLA adherence is not a major issue because the provider checks for locally available resources on resource reservation. If adequate resources corresponding to the specified SLA items are not available at reservation stage, the reservation request is rejected (partially or totally). Such rejection is handled at the federation level itself before the deployment request reaches a provider. However, the SLA enforcement for dynamic characteristics such as CPU usage should ideally be done by an SLA enforcement agent based on periodic monitored data as specified in the SLA.

2.3 Authorisation/Authentication Nightmare

Authorisation and authentication are required whenever a provider resource needs to be accessed. Primarily it is required to keep track of resource usage for billing and accounting purposes, but also for keeping computer and storage resources secure from malicious access. Private providers can easily maintain user accounts for authentication/ authorisation of their users. However, once part of a larger federation, it is not possible to keep account details of all possible users. The provider must support means of allowing resource access to federation users and at the same time keep at bay the malicious users.

One solution is to adopt a two-tier approach and keep the local authentication system intact for the non-federation users, and incorporate a federation authentication module in order to be able to validate the federation users' credentials before allowing access to local resources.

Another solution could be to elevate all local users as federation users, incorporate local accounts into the federation and provide all local users their new federation account details. But in this approach, the provider risks losing the business of local users to competing providers under the federation while gaining the simplicity of one unified authentication mechanism.

Regardless of which solution one adopts, there are bound to be headaches in the integration process with respect to users' authentication and authorisation with the (likely heterogeneous) providers in a large cloud federation.

The Contrail project will provide a uniform authentication and authorisation mechanism to its users' across all participating providers. The providers may maintain their own authentication mechanism for non-federation users. Adequate security mechanisms (Contrail consortium 2011c) will be in place in Contrail, the details of which are outside the scope of this chapter.

2.4 Issue of Trust in the Federation

A federation brings several disparate cloud providers under one umbrella. Every cloud provider has different guiding principles and motives. The issue of trust in the federation is very important for proper operation of the system. The cloud providers must trust the federation. They have certain expectations from the federation. They expect the federation to properly filter out malicious users, and to be honest in its accounting and billing practices. In a similar manner, the federation expects the participating cloud providers to adhere to a minimum code of conduct, be truthful in reporting resource usage, and properly honor any negotiated SLA contracts, etc. The federation and cloud providers can both maintain some sort of trustworthiness rating system. It need not be uniform across all participating parties, with each provider adopting (or not) any suitable rating system.

The Contrail project will incorporate a trust rating system based on a number of SLA violations, the nature of violations and other factors. The users will have an option to select a list of candidate providers for deployment of their applications (Contrail consortium 2011a). The trust rating system description is beyond the scope of this chapter.

3. User Interaction with a Contrail Federation

A federation can be considered as a bridge linking cloud users and cloud providers, with federation services acting as mediators between users and providers. From a user's perspective, the components and services supporting the federation act as a broker for the resources owned by providers participating in the federation. As each provider has its own, potentially different, mechanisms and policies for managing resources, the goal of the federation-support is to provide translation and mapping mechanisms for matching user needs and exploiting federated cloud providers.

Contrail will incorporate full support for OVF format for description and composition of distributed applications. Open Virtualization Format (http://www.dmtf.org/standards/ovf) is an open standard that is used to compose virtual applications. It allows for the inclusion of multiple VM templates, allows description of interconnection between VMs, and individual VM contextualization to name only a few features. It has been developed by the Distributed Management Task Force (DMTF, http://www.dmtf.org/), and is an up-and-coming standard with increasing vendor support.

3.1 REST APIs for Contrail

REST (Fielding 2000) stands for Representational State Transfer and is a simple yet elegant way to support API development. Using REST, one can provide simple CRUD—Create, Read, Update, and Delete—operations over identified resources. REST provides a very clean way of developing APIs and enables quick adoption from cloud developers. Since we desire Contrail modules including the VEP to be easily reusable in other projects, VEP will provide clear REST APIs for developers. In addition, VEP will have clearly defined RESTful APIs for other Contrail modules to communicate with. All requests from other modules including federation layer to VEP layer will be made using REST. The details of the APIs (Contrail consortium 2011a) are beyond the scope of this chapter.

The federation layer will also provide a REST interface in addition to a command line interface and a web interface for end users to interact with it. The REST interface that the federation layer in Contrail exposes to end users will enable its easy integration in third party applications, utilizing and in some cases extending the capabilities of the Contrail federation.

The REST APIs are built over Hyper Text Transfer Protocol (HTTP) protocol messages with support for HTTP GET/PUT/POST/DELETE messages only. Not all methods are available for all users and/or resources. Depending on the nature of the request and the access rights of users, a few or all methods are made available. The VEP REST interface, and the Contrail federation's external and internal REST interfaces will be stress-tested for scalability.

3.2 Basic User Scenario

In order to exploit a Contrail federation, the user first needs to register to the federation and get a user account. To interact with the federation, the user must be authenticated on the federation portal and obtain credentials (see fig. 1.2, step 1 below). They can then submit and monitor applications through the federation portal. The exact structure of the federation is transparent, as users never interact directly with the resource providers.

The basic interactions between a user and the portal are:

- **SLA negotiation:** the user negotiates the terms under which she can exploit resources from the federation: quantities (size and number of VMs, memory, processors, storage quotas), quality (performance guarantees), security (protection, localization), monitoring, reaction to events (elasticity management), pricing, etc. The results of SLA negotiation are a set of Constrained Execution Environments (CEE) that controls resource usage (see fig. 1.2, step 2 below).
- **Application submission**: the user interactively composes a new application from existing templates. The user can also submit her own applications using the OVF format (see fig. 1.2, step 3 below).
- **Application deployment**: the user provides the mapping between the application virtual resources and his CEE, which controls real resource usage
- **Application management and monitoring**: the user can monitor their application running on the federated cloud, check that the negotiated SLAs are correctly enforced and interact with their application (manage elasticity, take snapshots) (see fig. 1.2, steps 6 and 7 below).
- **Notifications:** the user is automatically notified of the major events from their application: failures, termination, SLA violation, etc. The events which will be notified can be negotiated in SLAs.

Fig. 1.2. A basic user interaction scenario with Contrail federation

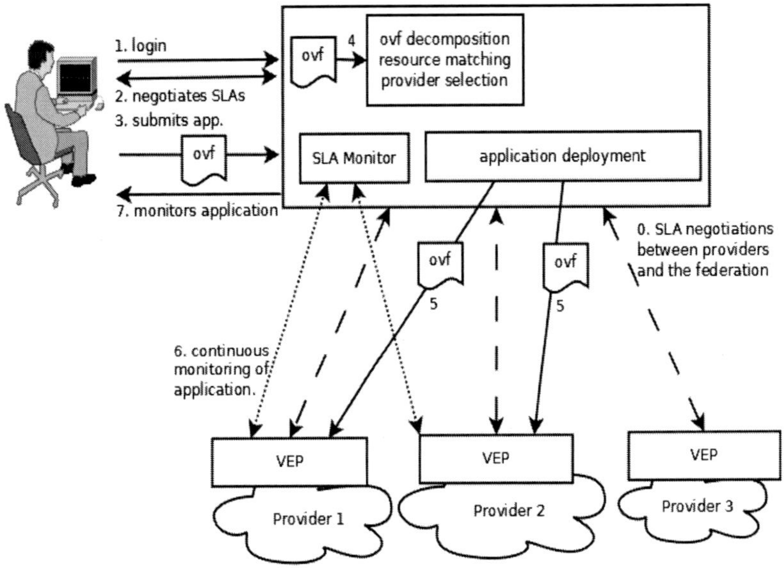

Contrail will also provide a means to submit applications for execution on a federated cloud in a single shot using a dedicated CTLapp document. This document contains all necessary information: SLA requests, OVF documents, application templates and deployment specifications.

3.3 Contrail Organisation

The Contrail software stack is organised in three major layers:

- **Federation layer**: the federation layer is in charge of user authentication, of negotiating SLAs with users and with providers, of finding resources and coordinating application deployment, of enforcing user negotiated SLAs, and of monitoring, accounting and billing for resource usage.
- **Provider layer**: the provider layer represents the business side of a provider. This layer is in charge of SLA negotiation with the federation, of the enforcement of these SLAs, of monitoring local resource usage and pushing monitoring data to the federation layer, and of accounting and billing for effective resource usage.

- **Resource layer**: this layer, called VEP in Contrail, is in charge of the exploitation of provider resources. It contains services for resource reservation, application deployment and monitoring.

3.4 User Authentication

In Contrail, users authenticate on the provider portal in order to exploit the federated cloud. Having a user registered with all providers in the federation would not be realistic, mainly because of the high number of providers. The federation layer dynamically manages user credentials on the providers when needed: when a provider is selected by the federation to run a user application, the federation first updates (pushes) the user credentials on this provider. These credentials can be destroyed when no more activity is planned for the user on the provider.

Contrail providers trust the federation for user management. For cases where it is not possible to establish trust between a provider and the federation, Contrail can use an alternative model: the user directly authenticates on the provider and gets their credentials. They can then delegate these credentials to the federation that will use them when interacting with this provider.

3.5 SLA Negotiation

SLA negotiation is handled in two steps on a Contrail federation: a user first submits an SLA request to the federation portal. The federation SLA manager can directly return an SLA proposition to the user if it has enough knowledge about the resource providers, or it can first try to negotiate the SLA terms with some providers before returning the proposition to the user. The SLA manager keeps track of the correspondence between the user SLAs and the SLA terms negotiated with different providers. A CEE is derived from each SLA negotiated between the federation and a provider and stored in the user descriptor in the provider structure. A CEE provides details about the terms negotiated in the SLA: resource type, quantities (memory size, number of cores, network throughput, etc.), quality (expected performance), non-functional terms (localisation), events and information to be monitored (periodicity, details) etc. Contrail SLA negotiation is based on the SLA@SOI (2011) framework.

3.6 Application Deployment

The deployment of a user application by the federation, based on OVF (http://www.dmtf.org/standards/ovf) in Contrail is coordinated from a Deployment Document (DD). For each item to be deployed (VM, network, storage), this document specifies the corresponding OVF item, one or more CEEs and any information necessary for the contextualisation of the item. This document can be directly provided by the user or automatically generated by the federation from the OVF and from the available user SLAs. Contrail also supports network aliasing: it is possible to request the mapping of a network item on the network of an existing application. In this case, both applications run on the same network and can interact.

The federation application coordinator service selects candidate resource providers from this document (SLAs corresponding to the specified CEEs must have been negotiated with the provider) and sends resource reservation requests to these providers. A reservation request can partially or totally fail if not enough resources are available. This process is iterated on available providers until all requested resources have been reserved. It is possible for the federation to negotiate SLAs on behalf of the user with extra providers during this process. From the list of selected providers, and the initial DD, the application coordination service derives one deployment document for each provider.

3.7 Multi-provider Application Deployment

A large-scale Contrail application can be distributed on resources from multiple providers. All these providers need not be selected when the application is initially deployed. It is possible that, on request from the user, from the application, or following some elasticity rule defined in an SLA, extra resources have been requested from the federation during the execution. It is even possible for providers to be selected later, although they were not part of the federation when the application was initially deployed.

To allow appliances to communicate, Contrail associates one (or more) Virtual Infrastructure Networks (VIN) to each application. The characteristics of the application VIN are defined by specific SLA terms: VIN type (private or public network), transport type (IPsec, VPN, etc.), level of protection, expected performance, connection process (contact point), etc. These terms are negotiated with the providers. A provider can inform the federation of network details such as the connection points (IP, port, credentials) in the SLA offer. It is the responsibility of the federation to ensure that the offers from various providers allow the building of the requested network. The network resources necessary to deploy VIN are reserved during the reservation step, prior to the deployment step. When the federation generates the individual DDs for each provider, extra contextualisation data gathered from SLAs or from reservation descriptors (contact points) are added by the coordination service to the network items to allow the setup of the VIN.

The appliances of the initial DD are distributed on the provider DDs. The network appliances augmented with contextualisation data are replicated in the individual DDs. Access to GAFS servers is also described in the DD in order to extend the VIN to these servers. It is possible that some virtual system (VM) defined in the OVF is deployed to a single provider. It would be possible to provide a modified OVF document containing only the items to be deployed to each provider. Such a strategy, however, would make elasticity management more complex as some items not deployed at the initial startup may later be necessary to extend the application. The solution adopted in Contrail is to replicate the OVF document on all selected providers and to deploy only the VMs listed in the deployment document.

3.8 Appliance Startup and Shutdown

Once the application has been deployed (all VMs in the "deployed" state), the execution can be initiated. It is enough to change the VM states in the DD to "running" to initiate the execution. Note that that setting the VM states to "running" in the initial DD will automatically initiate the startup once all VMs have been deployed. To start the application, the federation application coordinator follows the startup rules defined in the OVF document. These startup rules define startup steps with dependencies and possible delays between VM power-on. For each step, the application coordinator updates the individual provider DDs, changing the VM state which is to be started during this step to "running," submits the updated documents to the providers, waits for an updated status and then proceeds to the next step. The same process is applied for all state transitions (shutdown and suspend).

3.9 Resource Usage Monitoring

The VEP layer integrates a monitoring subsystem, in general based on a publish-subscribe system (see Section 4.5). The behaviour of this system is controlled by the SLAs associated to the resources. The SLA and the derived CEE define which information is to be monitored and at what frequency. The SLA enforcement service of the provider collects the monitoring data, checks for conformance and, if necessary, notifies other services (elasticity management) or the user.

In the context of a federation, appliances under control of the same SLA may have been distributed to multiple providers. All SLA terms, at least those that apply globally to all resources managed by the SLA—the total CPU usage per minute for instance—cannot be checked at provider-level. These SLAs must be enforced at federation level.

To effectively support SLA enforcement at federation level in Contrail, the SLAs negotiated between the federation and each provider will be enhanced with requests to forward monitoring data to the federation layer. The monitoring data of multiple appliances can be aggregated before being sent to the federation. To support this feature, the means to publish monitoring data to the federation layer is integrated in the provider layer.

3.10 Elasticity Management

Elastic applications support the dynamic provisioning and release of appliances during their execution in order to get more computation power during peak loads and to free resources when they are no longer required. Contrail supports elastic applications at the federation level as well as at the provider level. The management of elasticity in Contrail is based on updates to the deployment document. If a new VM item is added to the document, the corresponding appliance is added to the application. If an item is deleted, the corresponding resources are freed.

The deployment of new appliances follows the same steps as for the initial deployment. At the federation level, the resources for the new items are first reserved: the federation updates the individual deployment documents of the providers and submits them for reservation. The process can be repeated until the necessary resources are obtained. If not enough extra resources can be obtained from the currently selected providers, the federation can negotiate the necessary SLAs with extra providers, submit the OVF description and try to reserve resources.

Removing VMs from the DD automatically destroys the virtual machine. The application might not support brutal removal of appliances as this could be considered as a failure. Before suppressing the appliance, the VM can first be cleanly shut down and then removed from the DD when in the terminated state.

Elasticity can be automatically managed through SLAs or from application self-monitoring in Contrail, or manually on user request.

Elasticity rules can be defined in SLAs, based on data that can be monitored. These rules trigger registered actions when a certain threshold is reached. Monitoring the behaviour of an application from outside may be tricky unless the application is allowed to publish its internal data on the monitoring bus. Another solution is to allow the application to trigger predefined actions on the application coordination service of the federation. The application can, in this case, monitor its load, for instance the length of the task queue of a bag-of-tasks framework, request unused appliances to shutdown or request the application coordinator to deploy new appliances. The deployment documents are updated accordingly. Finally, the user can also update the DD at the federation level and so manually manage elasticity.

3.11 User Monitoring

The low-level Contrail monitoring system tracks the major events in the applications (start, stop, terminate, etc.) and periodically publishes data about resource usage. This data does not always allow the user to fully monitor their applications, such as how the VMs are distributed on the data centers and on the providers, how some non-functional constraints such as the localization are enforced. To provide full observability of their applications to users, the Contrail API provides the means to observe all details of their deployed applications: the distribution of the VMs on the physical resources, the associated CEEs and corresponding SLAs, the mapping of application OVF virtual systems on the VMs, etc. The Contrail RESTful API allows users to browse the details of their applications. Browsing is subject to access control. Access rights to the information can be managed by the owner (or manager) of the information using ACLs. For instance, a provider not willing to show which other VMs are running on their servers can restrict the user view to his own virtual machines. A user can also request for anonymity of their VMs during SLA negotiation.

3.12 Application Snapshot

To provide resilience to failures, Contrail supports application snapshots. In Contrail, a snapshot is a single OVF document describing all active VMs with links to their respective saved disk images.

To take a snapshot, the user just needs to update the deployment document. For each VM of his application, he can declare the state as "snapshot" and provide a path to the saved disk image. It is often not necessary to take snapshots of all VMs, for instance those providing stateless services. These VMs can be left in the "running" state, or moved to the "suspended" state. When the modified document is submitted to the federation, the application coordinator updates the individual provider DDs accordingly to reach the "suspended" state for all VMs. As for application startup, the federation application coordinator iteratively updates the individual DDs according to the OVF document.

When all VMs are in the "suspended" state, the state of the VMs for which a snapshot is requested is moved to "snapshot" in the provider DDs and the documents submitted. Each provider VEP is in charge of saving a copy of the VM disks in the "snapshot" state to the specified file storage path. In the meantime, the federation application coordinator generates a new OVF file. This new OVF is based on the initial one, with a new virtual system description added for each snapshotted VM in the same virtual hardware collection as the initial virtual system. The disk image section is updated accordingly to contain the paths to the disk image snapshots. It is possible to specify which initial virtual systems of the initial OVF are to be maintained in the new OVF when these virtual systems are required for elasticity management. An OVF snapshot can be submitted to the federation as a classical OVF application description. This OVF can be deployed on a new set of providers. The sole restriction is that the resulting OVF might be less adaptable than the original OVF, i.e. no alternative disk image, etc.

3.13 Virtual Organisation and Cloud Bursting

Contrail does not only support individual users but also user communities that can share resources on the cloud. Contrail users can manage access rights to their objects in the Contrail space using Access Control Lists (ACLs) based on their credentials. The exact description of user credentials (Contrail consortium 2011c) is beyond the scope of this chapter. To understand of the following text, we just need to consider that these credentials prove the membership of users to groups and their roles in the groups. A Virtual Organisation (VO) is a group of users. At least one of the users has credentials to create resources inside the VO—the VO admin.

The VO admin can negotiate SLAs and submit applications for execution on behalf of the VO. If allowed by the CEE ACLs, a VO user can reference a VO's CEE in the deployment document of their application. In this case, the corresponding virtual resource is controlled by the VO's CEE: accounting and billing are sent to the VO admin, etc. This allows, for instance, organisation users to exploit cloud resources on behalf of their organisation. Through the network aliasing capability, a user can deploy an application on a network deployed by the VO. This possibility allows a VO admin to create an Elastic Data Centre (EDC): the VO admin negotiates SLAs on a federation for various VM kinds, some storage and a network. The VO admin deploys the network and storage. As no VM is deployed, the resources consumed by this initial EDC are minimal. VO users can then deploy their own applications, mapping their application network on the VO network, and controlling their virtual machines by the VO CEEs. These VMs are added to the EDC and the VO is billed for resource usage.

Cloud bursting is implemented on a Contrail federation using VOs and EDCs. A real organisation managing its own private cloud using the Contrail stack negotiates VM SLAs and a network SLA with a Contrail federation (or directly with some cloud providers) and with its private cloud. The organisation then deploys the network, connected to its private cloud network. It should even be possible to configure this network to exploit IP addresses from the organisation Internet domain using VPN technologies. User deployment documents specify the mapping of their application network on the organisation network and the control of their VMs by the organisation CEEs. Applications submitted by the organisation users are deployed on the private cloud as long as resources are available. On peak loads (no more local resources available), the Contrail system requests resources from the public cloud. The VO network is automatically extended to the selected providers. The exploitation of external resources is transparent to the users as their applications run in the same context as if they were executed in the local cloud: same network, same services.

4. Virtual Execution Platform

With the growing importance of cloud-based services in the computing of tomorrow, it is only a matter of time before there will be a real need for merging disparate IaaS platforms in order to support increasingly complex applications of the future.

4.1 Need for VEP in Cloud Computing

The Virtual Execution Platform (VEP) could provide that aggregating glue in bringing different cloud platforms under one umbrella. The standard APIs exported by the VEP software would allow other developers to quickly and easily develop tools for making their own federated platforms tuned for specific tasks that other federations including Contrail may not support.

There are several open source tools already available for accessing public clouds (Amazon EC2), and integrating such tools with VEP would allow for quick and easy solutions in bringing public and private clouds together. Thus, a well-designed VEP solution could usher in the next wave of cloud research and open up new vistas for cloud innovations. Keeping this motivation for developing a quality VEP, let us next look at Contrail's VEP design in detail.

4.2 Contrail Virtual Execution Platform

Virtual Execution Platform (VEP) is a Contrail service that sits just above IaaS layer at the service provider end of the Contrail cloud federation. Some major design goals were kept in mind while designing the VEP architecture. Support for open virtualisation standards including Open Virtualisation Format (OVF) has been one of the major requirements. To enable application distribution on multiple providers, partial OVF deployment specified in a deployment document is also supported. Scalability of the API interface in order to support a potentially large customer base, interoperability with various open IaaS technologies in order to bring several kinds of cloud providers under the Contrail umbrella, and providing support for elasticity in accordance with the negotiated SLAs have been other major requirements that VEP will satisfy. Ease of deployment is another requirement; the VEP software will be distributed as a self-contained software package. Cloud providers will be required to just install it and configure a few parameters that will enable VEP to properly interface with their installed IaaS platforms. There will be no major modification that will be required in the IaaS layer itself. VEP will contain drivers for interfacing with various IaaS layers using (well known) provider APIs.

It will have selectable built in drivers for various open source IaaS platforms such as OpenNebula and OpenStack. The VEP layer will integrate clouds composed using different technologies under the ambit of Contrail federation. VEP will be very closely integrated with the IaaS layer and will expose uniform APIs to higher layers in the Contrail architecture. It will enable seamless deployment of OVF applications over any underlying supported cloud platform.

In case the federation module suspects that none of the participating Contrail cloud providers can satisfy a user's application resource requirements, it may enable part-deployment of the appliance over public clouds such as Amazon EC2 (cloud-bursting) utlising user credentials for such public clouds. This capability will be handled by the federation layer and not VEP directly.

Since support for Service Level Agreements (SLAs) and providing Quality of Protection (QoP) to end users is a major feature that Contrail aims to provide, the VEP layer has also been designed keeping the larger Contrail requirements in the picture. The following subsections describe major VEP components and supported features in some details.

4.3 VEP's OVF Deployment Subsystem

Contrail VEP will provide full support for OVF v 1.1 standards (Crosby et al. 2010). The support for OVF in VEP will enable deployment of virtual applications described in OVF XML documents over supported IaaS platforms regardless of their native support for OVFs.

The OVF deployment subsystem in VEP will be responsible for deploying user appliances described within the OVF XML document. Fig. 1.3 shows the simplified design of the VEP's OVF deployment module. The user's OVF file is received from the federation layer using REST protocol (refer to fig. 1.3, subcomponent a). The user's credentials and authentication parameters are sent as part of HTTP headers in the Contrail REST scheme. Once the request for OVF deployment has been verified, the OVF document along with the corresponding SLA object is sent to the OVF deployment subsystem of VEP stack (fig. 1.3, subcomponent b).

Once the OVF document and the corresponding SLA object are received at the VEP layer, the OVF document is validated for correctness against DMTF OVF standard schema document (currently VEP supports OVF v 1.1—see fig. 1.3, subcomponent c). If the OVF document fails the validation process, the deployment effort is terminated and a suitable response is reported to the federation layer.

Upon successful validation, the OVF document is parsed (fig. 1.3, subcomponent d) and individual elements deconstructed. The deconstructed OVF components, including VM elements, network elements etc., are then linked to their corresponding SLA contracts deconstructed from the SLA object. These linked elements are then used as input in the IaaS translation function which, depending on the type of IaaS environment, generates the template/deployment files for the underlying IaaS platform. For the initial release, VEP aims to provide support for OpenNebula IaaS clouds (fig. 1.3, subcomponent f). Later on support for other open source cloud technologies will be developed.

Fig. 1.3. OVF subsystem in Contrail VEP (simplified version)

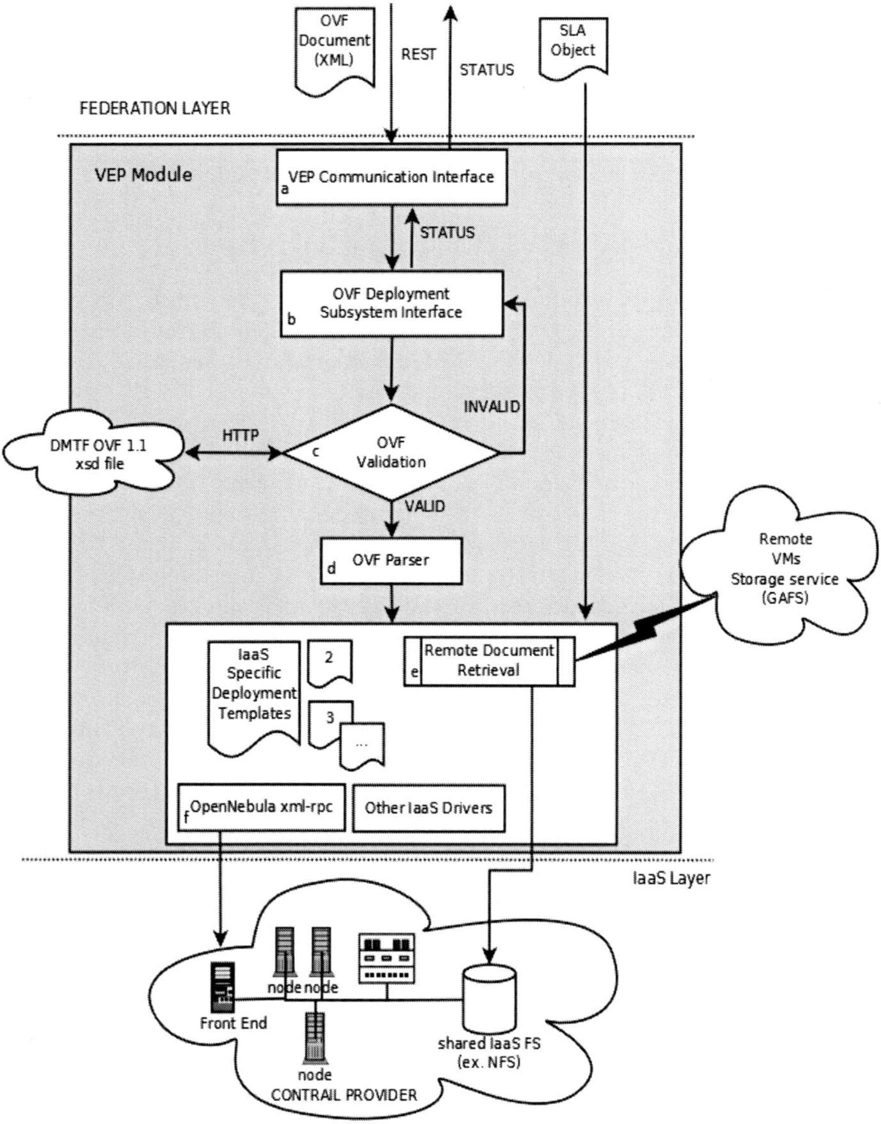

If a virtual machine is stored at a remote location, the OVF deployment subsystem has modules for authenticated remote file transfer (fig. 1.3, subcomponent e). The remotely hosted resources (VMs) will be transferred onto the cloud provider's shared file system before the actual virtual appliance deployment takes place at the cloud provider. The Contrail project will incorporate a Global Autonomous File System (GAFS) based on the XtreemOS project (Morin 2007). The users' virtual machine image files will be hosted at remote GAFS shares and the OVF subsystem will log on behalf of the user and transfer image files from remote GAFS locations to the cloud provider shared file system.

4.4 OCCI Support

The Open Cloud Computing Interface (OCCI, http://occi-wg.org/) is a set of open standards (http://occi-wg.org/about/specification/) for managing cloud resources developed through the Open Grid Forum (OGF, http://www.ogf.org/). The VEP component in the Contrail project will have an extended OCCI support. Again, the need to incorporate OCCI in VEP arises from the requirement to be extensible and reusable in other projects. Additionally, as a result of our work and the experiences gained in providing support for OVFs and SLAs, the Contrail project will propose an extension to the current OCCI draft for inclusion in standardisation within the OCCI community. These extensions will allow the OCCI framework to support OVF deployments along with providing support for SLAs that the standard lacks in its current format. The support for OCCI in VEP will hopefully allow the platform to be integrated in future cloud projects and will allow the open source community to develop feature enhancements and add-ons to the VEP layer in the federation of clouds.

4.5 Application Monitoring and Reporting Module

Application monitoring subsystem is a very integral part of the VEP as it provides support for the proper enforcement of the SLA contracts. Every application that is deployed on a provider's IaaS infrastructure is monitored by the VEP module periodically. The collected data is stored in the VEP database for on-demand retrieval on the user's behalf by the federation, but few statistics are pushed periodically by the monitoring bus to be used by the SLA enforcement module to check for compliance and for making elasticity decisions. Fig. 1.4 shows the simplified schematic of the monitoring subsystem in the VEP module.

Fig. 1.4. Reporting subsystem in Contrail VEP (simplified version)

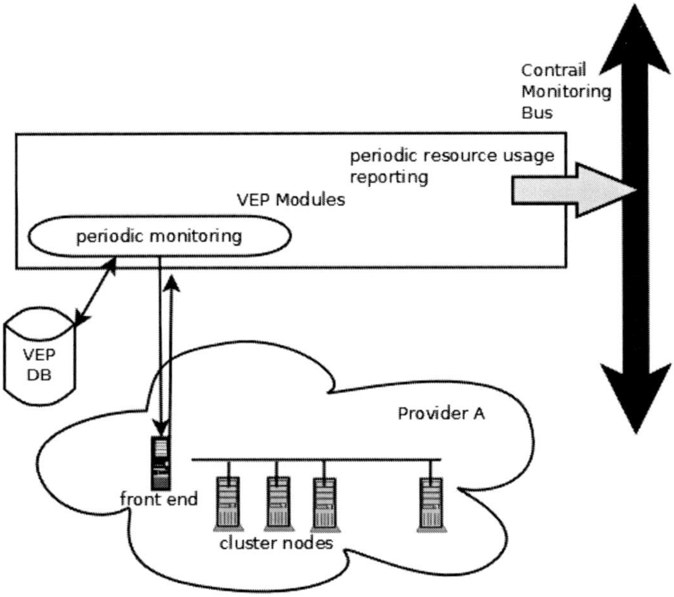

Each deployed application is assigned a Universally Unique Identity (UUID) by the Contrail federation. The same UUID is included as part of the reported data that helps various modules in the SLA/monitoring system in the federation link the data with the correct application and the negotiated SLA object for verification. The federation monitoring bus over which the periodic data is sent will be a distributed messaging system (example: Apache ActiveMQ, http://activemq.apache. org/) that supports Java Messaging Service (JMS) APIs.

4.6 IaaS Drivers

The VEP software is envisioned for use with multiple open source IaaS clouds. Contrail VEP software eventually plans to support several major IaaS platforms including OpenNebula, OpenStack, Nimbus scientific cloud, etc., and appropriate drivers that can be enabled by the cloud providers as desired for interfacing with appropriate IaaS platforms will be provided.

4.7 Federation Authentication / Authorisation Support

The VEP layer helps a cloud provider become part of the larger Contrail cloud federation. The VEP layer handles all communication to and from the cloud provider and the federation. All requests on behalf of the federation user to the provider come from the federation layer. Each such request has to be properly authenticated and only authorised requests should be forwarded to the IaaS layer. This check is necessary to prevent resource abuse and is also required for accounting and billing purposes.

Since VEP communicates to the federation through REST APIs, each request carries with it the user authorisation credentials embedded in the HTTP headers. User certificates and a trusted chain of certificates are used for verification of authorisation credentials. The federation maintains the registered user accounts and the resource access rights and other details. The VEP is only interested in finding whether the presented credentials are authentic or not. Once the credentials are verified, the provider's resource access request is forwarded to the proper IaaS cluster.

The VEP also maintains a temporary access control table for active users and their applications that have been deployed through itself. An additional check against this table may also be performed if deemed necessary against the type of access requested.

4.8 Virtual Organisation

The Contrail VEP supports the notion of a Virtual Organisation (VO). The cloud provider/administrator will have the ability to configure the parameters of the virtual organisation including list of hosts, networking connections, storage. The VO users' can submit OVFs for deployment, making use of already provisioned VO resources. The billing for such resources will be sent to the VO and not the individual VO users. The VEP will support VOs even if the underlying IaaS platform does not support such a notion of organisation. If a supported IaaS platform supports VOs natively (for example OpenStack), VEP will make an effort to utilise the supported feature as much as feasible. Thus, VEP will provide a uniform notion of VO to end users regardless of the underlying IaaS support of VOs.

5. Concluding Remarks

This chapter has presented a detailed design of Contrail VEP implementation and the rationale for the design. We have discussed the Contrail cloud federation briefly while exploring the major challenges any cloud provider would face and must address if they plan to become part of a larger federation. The proposed Contrail VEP software support for a broad set of open cloud platforms would help a significant number of private cloud operators to seamlessly join a Contrail federation. Furthermore, VEP's support of open standards and a well defined API set will enable cloud researchers and developers to easily extend its features to suit their future requirements, thus making VEP somewhat future proof.

VEP's fully integrated support for Contrail federation authentication and authorisation mechanisms, and a fully stress-tested REST and OCCI interface will address some of the challenges outlined (see Section 2.3 above) for being part of a federation. The open source mindset and access to the source code would allow independent verification of monitoring modules and other features and thus would help build some level of trust in VEP and Contrail federation (see Section 2.4).

The Contrail project road map is well-defined; in the initial release the VEP effort will provide complete integration with the OpenNebula IaaS platform. Support for other cloud platforms including OpenStack will follow in the subsequent releases. There will be basic scheduling support in the first release, but a full resource scheduling feature in the second release is planned. A full support for OCCI will also be provided in subsequent releases. The first release will also see basic OVF support with more complete support in subsequent releases.

Acknowledgements

This work has been funded by the European Commission through the Contrail project, contract number 257438, within the 7th Framework Program.

References

Contrail Consortium. "Architecture Design and QoS constraints matching algorithms in Federations." August 2011. Contrail Deliverable D2.2, 2011a.

—. "Requirements on Federation Management, Identity and Policy Management in Federations." March 2011. Contrail Deliverable D2.1, 2011b.

—. "Security Requirements, Specification and Architecture for Virtual Infrastructures." March 2011. Contrail Deliverable – D7.1, 2011c.

—. "Requirements and Architecture for the Global Autonomous File System." March 2011. Contrail Deliverable D6.1, 2011d.

Crosby Simon, Doyle Ron, Gering Mike et al. "Open Virtualization Format Specification." DMTF, 1.1.0 edition, 2010.

Fielding Roy Thomas. "Architectural styles and the design of network-based software architectures." PhD thesis, AAI9980887, 2000.

Morin Christine. "XtreemOS: A grid operating system making your computer ready for participating in virtual organizations". In Proceedings of the 10th IEEE International Symposium on Object and Component-Oriented Real-Time Distributed Computing, ISORC 2007.

SLA@SOI Consortium 2011. "SLA Model". http://sourceforge.net/apps/trac/sla-at-soi/wiki/SLA\%20Model\%20\%3A\%20The\%20Model (accessed September 01, 2011)

Sotomayor Borja, Keahey Kate, and Foster Ian. "Combining batch execution and leasing using virtual machines." In Proceedings of the 17th international symposium on High performance distributed computing, HPDC, ACM, New York 2008.

CHAPTER TWO

TOWARDS AUTONOMIC TRANSACTIONAL REPLICATION FOR CLOUD ENVIRONMENTS

MARIA COUCEIRO, PAOLO ROMANO AND LUÍS RODRIGUES

In recent years, in-memory transactional data platforms, often referred to as NoSQL data grids, have emerged as a reference solution for transactional data management in the cloud. In these in-memory platforms, replication plays a role of paramount importance for fault-tolerance purposes, given that it represents the key means to ensure data durability in the face of unavoidable node failures. Unfortunately, despite the abundance of approaches in the area, existing replication schemes still fall short of addressing one of the key requirements of cloud computing environments: ensuring optimal efficiency when deployed over elastic infrastructures that dynamically expand/reduce the number of (physical or virtualised) resources in response to fluctuations of workload characteristics.

Tackling this issue is a key goal of the Cloud-TM project, a recent EU funded initiative that aims at developing a self-tuning, elastic transactional data platform tailored to meet the elasticity requirements of cloud computing infrastructures.

1. Introduction

By shifting data and computation away from local servers towards very large scale, world-wide data centers, cloud computing promises compelling benefits for both cloud consumers and cloud services providers (Armbrust et al. 2009). However, the promise of infinite scalability catalysing much of the recent hype about cloud computing is still menaced by one major pitfall: the lack of programming paradigms and abstractions capable of bringing the power of parallel programming into the hands of ordinary programmers.

One of the most crucial issues to tackle when developing applications for the cloud (and in general for large scale distributed systems) is the management of concurrent access to a shared state. The challenge here is to identify mechanisms capable of ensuring adequate consistency levels while being:

(1) simple and familiar for the programmers

(2) highly efficient and scalable

(3) fault-tolerant and highly available

Cloud-TM (www.cloud-tm.eu) is an ongoing European Project that is tackling precisely these issues by developing a self-optimising middleware platform aimed at simplifying the development and administration of applications deployed on large scale cloud computing infrastructures.

At the core of the Cloud-TM platform lies the abstraction of a Distributed Software Transactional Memory (DSTM) (Carvalho et al. 2010; Couceiro et al. 2009). DSTMs are a recently proposed extension of the Transactional Memory (TM) (Shavit and Touitou 1997) programming paradigm, which was originally introduced to simplify the development of concurrent, though not distributed, programs. By releasing the programmer from the burden of managing locks or other error-prone low-level concurrency control mechanisms, TMs have been shown to enable a significant boost in productivity, to shorten development times, and to increase code reliability in complex concurrent applications. DSTMs have recently garnered significant interest as a more flexible and scalable alternative to conventional relational DBMSs (Stonebraker et al. 2007), particularly attractive in cloud environments. On the industrial side, the market of in-memory transactional data grids (as DSTMs are typically referred to in these enviroments) has undergone a rapid proliferation, and nowadays offers a good number of platforms both as proprietary products (e.g. Oracle© Coherence) and open source projects (e.g. JBoss© Infinispan). On the academy side, this area is attracting the interest of a growing community of researchers from areas such as distributed computing theory (Attiya et al.2010), database systems (Kallman et al. 2008), high performance computing (Bocchino et al. 2008) and, naturally, transactional memories (Herlihy and Sun 2007).

In these in-memory platforms, replication plays a role of paramount importance for fault-tolerance purposes, since it represents the key means to ensure data durability in the face of node failures. Unsurprisingly, the replication schemes employed in these platforms take inspiration from the vast amount of literature on replication of transactional systems (Patino-

Martínez et al. 2000, Couceiro et al. 2009). However, despite the abundance of existing replication strategies, no universal, one-size-fits-all solution exists providing optimal performance in highly heterogeneous/ fluctuating workloads and independently of the scale of the underlying platform.

The complexity of this problem is therefore exacerbated in cloud computing platforms due to the feature regarded as one of the key advantages of the cloud: its ability to elastically acquire or release resources, de facto dynamically varying the scale of the platform in real-time to meet the demands of varying workloads.

In light of these considerations, we advocate that, in order to match the elasticity requirements of cloud computing infrastructures, and to maximize efficiency, in-memory transactional data grids should entail multi-modal replication strategies and be able to dynamically self-tune their operating mode.

The structure of this chapter is as follows: first, in Section 2, by surveying existing approaches for state consistency in replicated transactional platforms, a taxonomy of the main solutions in literature is presented. In Section 3, we complement the critical analysis with a set of performance evaluation studies aimed at exposing advantages and drawbacks of existing solutions. Our results provide experimental evidence of the fact that fluctuations of the workload can have a strong impact on the performance of existing replication protocols, motivating the need for investigating self-tuning mechanisms. In Section 4, an case study highlighting the benefits achievable via the proposed self-tuning replication mechanism is illustrated. Finally, Section 5 presents an overview of the online research lines, shedding light on some of the key challenges that need to be addressed to achieve this ambitious goal, and providing pointers to recent results in this area.

2. A Taxonomy of Transactional Replication Protocols

Over the last decades, a vast amount of literature on replication of transactional systems has emerged (Pedone et al. 2003; Kemme and Alonso 1998; Patino-Martínez et al. 2000). The various solutions can be classified according to the following taxonomy in fig. 2.1.

A first classification parameter is whether write transactions can be executed on a single node or by all nodes in the system: single versus multi-master schemes.

Fig. 2.1. Taxonomy for full-replication transactional solutions

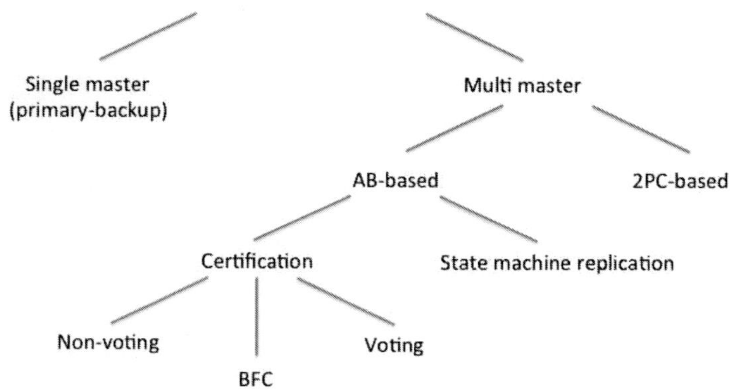

Single-master. In single master schemes, also known as primary backup, write transactions are executed exclusively at a single node (also called master or primary), whereas the remaining replicas can only run read-only transactions. Upon failure of the master, a backup replica is elected to become the new master.

Fig. 2.2 below exemplifies the execution of a transaction is a system with two backup replicas. The master executes transaction T1 and, after committing it, broadcasts its updates to the replicas, which apply them in order.

Note that, as the write transactions can be serialised locally by the master using its local concurrency control algorithm, this approach can rely on a simple replica synchronisation scheme with respect to multi-master solutions (as we will see shortly).

Fig. 2.2. Single-Master

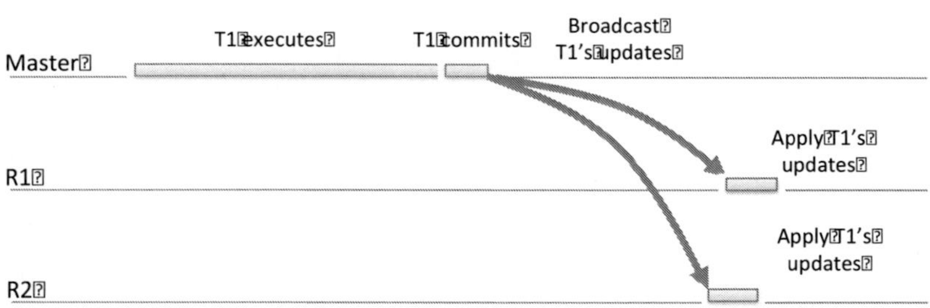

On the down side, the throughput of write transactions does not clearly scale up with the number of nodes in the system, which makes the master prone to becoming the system bottleneck.

Multi-master. Multi-master schemes, on the other hand, are typically more scalable as transactions can be processed in all nodes. There are two types of synchronising the access to data: eager and lazy. The first relies on a remote synchronisation phase upon each (read/write) access, which normally results in very poor performance results (Franklin et al. 1997). Conversely, the lazy approach defers replica synchronisation until the commit time, which is when the transaction is finally validated. Lazy multi-master schemes can be classified based on whether they rely on Atomic Commit Protocols (such as Two-Phase Commit) or Atomic Broadcast (AB) (Defago et al. 2004) schemes to determine the global serialisation order of transactions.

Two-Phase Commit. In solutions based on Two-Phase Commit (2PC), transactions attempt to atomically acquire locks in all nodes. Even though they normally incur in minor communication overheads with respect to those relying on AB, these solutions are well known to suffer scalability problems due to the rapid growth of the distributed deadlock rate as the number of replicas in the system grows (Gray et al. 1996).

Fig. 2.3. The Two-Phase Commit protocol

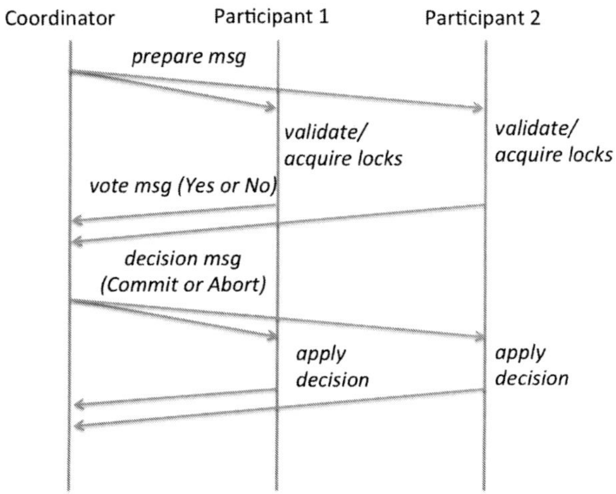

Fig. 2.3 above shows the 2PC protocol in action. The coordinator first sends the prepare message with the transaction. Both participants validate the transaction, acquire the locks of the objects read and written and reply to the coordinator with "commit" or "abort." If all participants answer commit, the coordinator sends a decision message with the indication of committing the transactions. If at least one participant replies abort, the coordinator sends an abort message. The participants apply the decision of the coordinator and the latter ends the protocols after receiving the acknowledgement from all participants.

Atomic Broadcast based replication. Conversely, AB-based replication is a family of (distributed) deadlock-free algorithms that serialise transactions in the total order established by an AB service (Defago et al. 2004). These solutions can be distinguished into two further classes: state machine replication and certification.

State Machine Replication. In the state machine replication (Schneider 1993), all replicas execute the same set of transactions in the same order. The transactions are shipped to all replicas using total order broadcast and, consequently, all replicas receive transactions in the same order and execute them in that order. However, both transactions and the validation scheme must be fully deterministic so that all replicas begin and end transactions in the same state.

Fig. 2.4 depicts the execution of the State Machine Replication in two replicas. Replicas 1 and 2 atomic broadcast the request to start transactions T1 and T2. T1 is ordered first—it (atomically) pre-acquires all the locks it requires, executes and commits in both replicas. T2 must wait until the locks are released to be executed also in both replicas.

Fig. 2.4. State Machine Replication

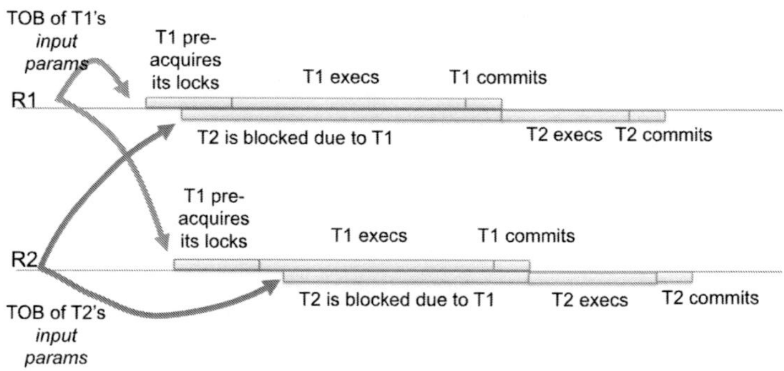

This solution has the key advantage of never incurring in transactional aborts. On the other hand, it requires a-priori knowledge of the transaction's read-set and write-set. Further, the execution cost of write transactions has to be fully paid by all the replicas in the system, which may limit the scalability of this approach in write dominated workloads.

Certification. Certification based techniques take a more optimistic approach and have been shown to achieve higher scalability by fully executing the transaction only at one node. This means that different transactions may be executed on different replicas concurrently. If the transaction aborts, no coordination is required. However, if the transaction is ready to commit, coordination will both ensure serialisability (Bernstein et al. 1986) and propagate the updates to the other replicas. This means that two transactions may update concurrently the same data in different replicas and it is up to the coordination phase to detect this situation and abort at least one of the transactions.

In this work, we will be mainly focusing on the following three certification protocols: non-voting (NVC), voting (VC) and Bloom filter (BFC) certification.

Non-Voting Certification. As with all certification protocols, in Non-Voting Certification (Pedone et al. 2003; Patino-Martínez et al. 2000) the transaction first executes locally. Then, when it is ready to commit, its read and write-set are sent to all replicas using Atomic Broadcast. Consequently, all replicas certify the transaction in total order. The validation phase consists of checking if the read-set is still valid or, in other words, if no other transaction has updated any item in the read-set. If the read-set is valid, the transaction can be committed. Otherwise, it is aborted.

Fig. 2.5. Non-Voting Certification protocol

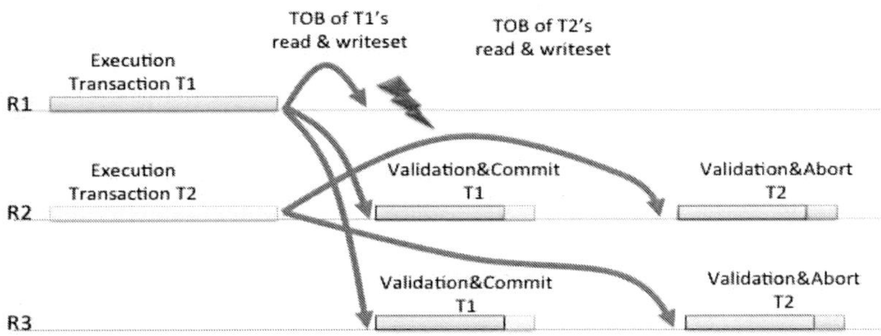

Fig. 2.5 above exemplifies the execution of NVC when one replica fails. Two transactions, T1 and T2, execute concurrently in two replicas. Atomic broadcast orders T1 before T2. R1 fails, but since the remaining replicas both received T1's read and write sets, they validate the transaction and apply the updates. T2 is then delivered to R2 and R3, validated and aborted due to conflicts with the already committed T1.

The key advantage of NVC consists of necessitating a single AB-based communication phase to determine the outcome of transactions. However, since the entire read-set must be sent together with the write set, messages exchanged by the replicas become very large, resulting in high levels of network congestion.

Voting Certification. The Voting Certification (Kemme and Alonso 2000) tries to overcome this issue by only ABcasting the write-set of locally executed transactions. As fig. 2.6 shows, commit requests are processed in total order but, in this case, only the replica where the transaction was executed can validate it since no other replica can access its read-set. When this replica finishes the validation, it sends the outcome to all replicas using reliable broadcast (Guerraoui and Rodrigues 2006), a lightweight communication primitive that guarantees all replicas receive the message. When this message is received, replicas either apply the write set of the corresponding transaction or discard it.

The messages exchanged by the replicas are significantly smaller compared to those generated by NVC. However, this protocol requires one additional communication step to disseminate the outcome of the transaction.

Fig. 2.6. Voting Certification protocol

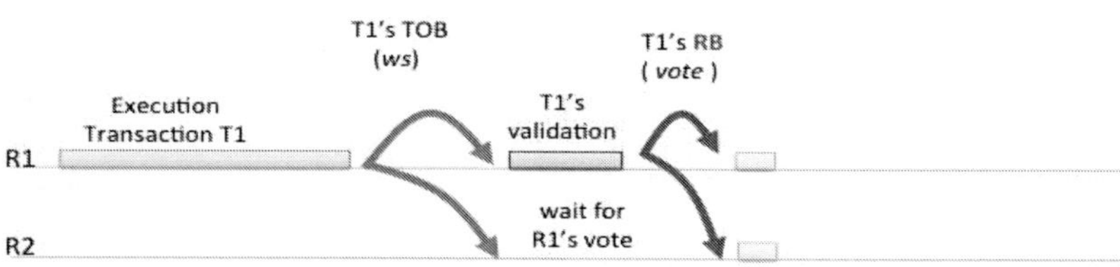

Bloom Filter Certification. Bloom Filter Certification (Couceiro et al. 2009) consists of encoding the read-set of transaction in a Bloom Filter (Bloom 1970), a space-efficient data structure that strongly (and efficiently) compresses the messages disseminated via the AB service, while still allowing every replica in the system to deterministically certify the transactions. Unlike voting schemes, BFC avoids additional communication steps during the commit phase. In terms of generated network traffic, even though BFC generates larger messages than voting protocols, it typically reduces the size of the messages exchanged via the AB service significantly when compared to non-voting schemes. On the down side, BFC can suffer from false positives due the probabilistic nature of Bloom filter-based encoding, which ultimately lead to an additional rate of aborted transactions.

3. No "One-Size-Fits-All" Solution

We can conclude from the previous section that the design space of distributed state consistency mechanisms is so vast that no universal, one-size-fits-all solution exists. The efficiency of individual state management approaches can in fact be strongly affected by both:

(1) the characteristics of the incoming workload, such as the ratio of read/write operations, as well as the spatial/temporal locality in the data access patterns, and

(2) the scale of the system (e.g. low vs high number of nodes, local vs geographical distribution) on which these mechanisms are deployed.

In order to support this claim, in the remainder of this section we present an experimental study to compare the performance of the previously presented protocols in distinct workload/scale settings.

The tests were run on a cluster of eight nodes, each one equipped with two Intel Quad-Core XEON at 2.0 GHz, 8 GB of RAM, running Linux 2.6.32-26-server and interconnected via a private Gigabit Ethernet.

In the two first experiments, the test system consisted of a key/value store implemented in Java. We used a synthetic benchmark and two different contention scenarios: the first consisted of a data set with 100,000 items, while in the second the data set had only 1,000 items. Note that by varying the size of the data set, the probability of contention among transactions also varies. Specifically, the contention probability is around two orders of magnitude larger when using the smaller data set. Each transaction performed ten reads and one update operation, choosing the data items to be accessed uniformly from the data set.

Single versus Multi-master. The graph in fig. 2.7 compares the throughput (number of transactions per second) of single-master (SM) and an implementation of multi-master using the 2PC protocol (MM-2PC) that is run at the end of each transaction. This experiment clearly showed that neither solution outperforms the other in both scenarios. For the low contention scenario, the MM-2PC is able to achieve a much higher throughput than single-master, whereas when there is high contention in the data set, the single-master scheme outperforms MM- 2PC. This is due to the fact that, while single-master is unaffected by the contention, the number of conflicts has a huge impact in the performance of MM-2PC, and the low throughput is the result of the high number of deadlocks.

Fig. 2.7. Throughput of single-master and multi-master 2-Phase Commit protocols

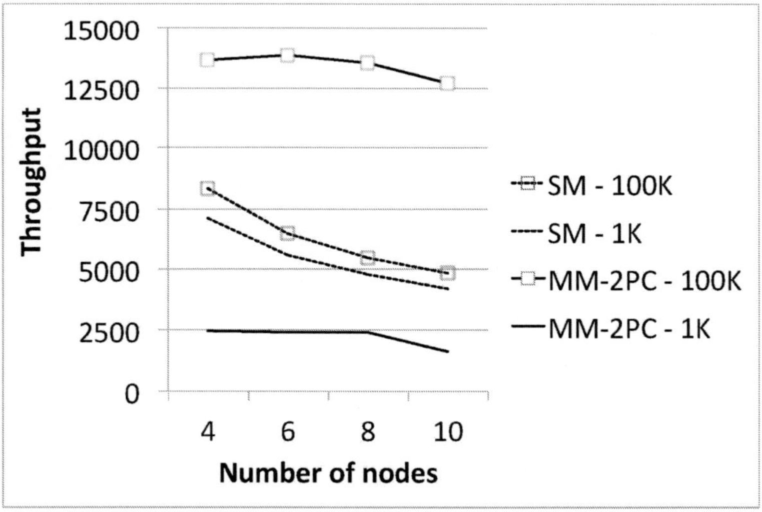

Multi-master versus Certification. The plot in fig. 2.8 compares the performance of MM-2PC and certification. The MM-2PC performance has the best performance for the low contention scenario. Certification outperforms MM-2PC in high contention scenarios once more due to the high rate of deadlocks in the latter.

There are many approaches to implement Atomic Broadcast (Defago et al. 2004), for instance sequencer (one node is responsible for ordering the messages) or token (nodes can only send messages when they own a token). For this test we used a sequencer based implementation of Atomic Broadcast which limits the maximum throughput achievable by the system to the maximum number of messages that can be ordered by the sequencer. This is the only bottleneck in the performance of the protocol, as the contention level does not influence the throughput.

Fig. 2.8. Throughput of Certification and multi-master 2-Phase Commit protocols

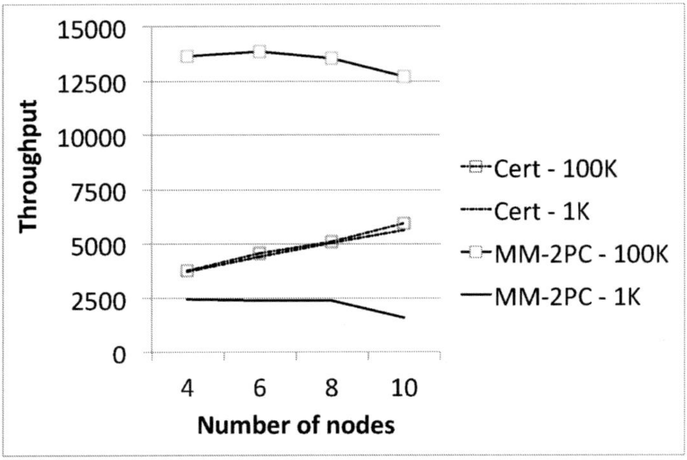

Non-Voting versus Bloom Filter versus Voting Certification. Following the discussion presented in Section 2, the performance of each of these three alternative certification mechanisms is strongly dependent on the actual distribution of the size of the read-sets generated by the transactional application.

Fig. 2.9. Throughput of three certification strategies with different read-set sizes

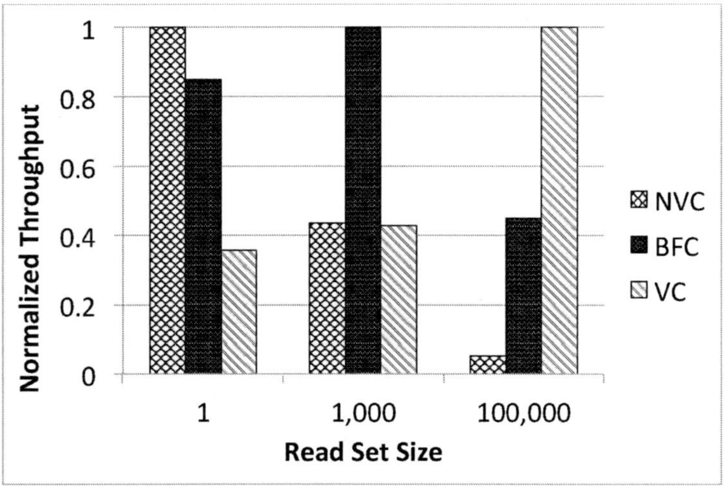

Fig. 2.9 shows the results of a sensitivity analysis aimed at assessing the actual impact of the read-set size distribution on the performance of the three previously described certification schemes. The results were obtained using a simple synthetic benchmark adapted from the Bank Benchmark originally used in Couceiro et al. (2009). Focusing only on the effects due to variations of the read-set size, which represents the goal of this sensitivity analysis, the benchmark was configured to never generate conflicts among transactions. The only aborts experienced in the system are therefore those determined by false positives with the BFC scheme (which was configured to have an additional abort-rate of 1%, as in Couceiro et al. 2009). The in-memory transactional data grid and the certification protocols were implemented in Java. The system uses two main components: (i) a state-of-the-art Software Transactional Memory (STM), namely JVSTM (Cachopo and Rito-Silva 2006), used to manage local concurrency, and (ii) a replicated key/value store, used to maintain associations between unique object identifiers and object instances.

Our experimental results highlight that no one-size-fits-all-solution exists that maximizes throughput across all the considered workloads. On the contrary, NVC provides the best performance in the scenario with small read-sets, BFC the best in the scenario with 1,000 items in the read-set, and VC is by far the best with large read-sets. Further, the relative difference in the performance between the best and worst performing protocol for each scenario ranges from a factor 2.5x (BFC vs VC, read-set size equal to 1,000) to 10x (VC vs NVC, read-set size equal to 100,000).

4. The Cloud-TM Approach

Section 3 provided a number of experimental evidences highlighting the benefits that could be achieved by dynamically adapting the replication mechanism in an in-memory transactional data platform. Fig. 2.10 below illustrates a possible example scenario highlighting how the autonomic, self-optimising capabilities that we are developing in the Cloud-TM platform will be exploited by user level applications.

The bottom part of the figure shows the variations over time of (a) the incoming traffic, (b) the ratio of read/write transactions, and (c) the transaction conflict probability.

Five different workload scenarios are considered, each associated with distinct optimal management strategies minimising the number of resources to be hired from the cloud (while, of course, ensuring the same, user-defined Service Level Agreements).

Fig. 2.10. Example scenario of self-tuning replication in the Cloud-TM platform

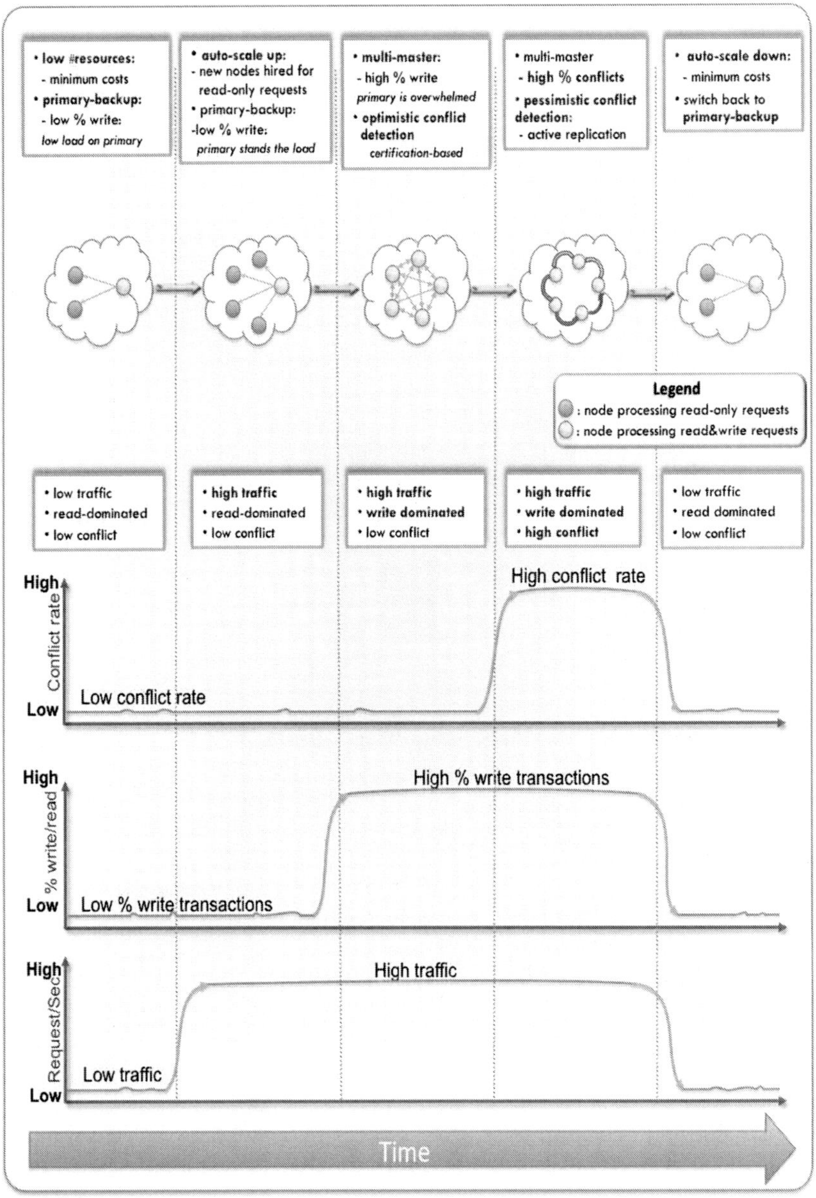

In the first scenario (from the left) the workload is read-dominated and both the incoming traffic and the transaction conflict probability are low. In these settings, the optimal strategy is to hire a low number of resources and to use a primary-backup replication scheme. As discussed in Section 2, this simple, centralised approach allows the minimising of the replica coordination overhead, providing optimal performances as long as the number of write transactions remains sufficiently low so to avoid overloading the primary.

As the load increases (second scenario from the left), leaving the conflict rate and the write/read ratio unaltered, the Cloud-TM will automatically scale up the number of nodes hired from the cloud infrastructure. The replication scheme does not vary with respect to the former scenario. In fact, being workload read-dominated, the primary is still able to quickly process the incoming write transactions.

In the following scenario (the third from the left), due to a surge in the relative number of write transactions, the primary gets overwhelmed by the incoming write requests, becoming the system's bottleneck. The Cloud-TM will react by triggering a reconfiguration of the replication scheme, passing to employ a multi-master replication scheme which allows for an even distribution of the write requests among the nodes and, consequently, the achievement of a higher throughput. Since the transaction conflict rate remains low (because the write operations target low contended data regions, for example private user data), an optimistic conflict detection policy relying on a certification-based approach is adopted. In fact, by letting transactions optimistically access possibly stale data, this class of replication protocols allows for achieving maximum inter-replica parallelism, at least as long as the conflict rate remains sufficiently low (Ciciani et al. 1992).

In the fourth scenario, we consider precisely the increase of the transaction conflict rate. This type of workload may arise, for instance, in an e-auction application in proximity to the expiration of an auction for a highly contended item. As already hinted, in high-contention scenarios, optimistic conflict detection schemes suffer from a high abort rate and incur a significant performance degradation. The Cloud-TM middleware will detect such a suboptimal system configuration, and transparently switch to a more pessimistic conflict detection scheme, such as the active replication mechanism described in Section 2, which is known to perform better in high conflict scenarios by serialising transactions and reducing the transaction abort probability (Ciciani et al. 1992; Nicola and Jarke 2000).

Finally, to the extreme right, as the workload returns to exhibit its original characteristics (low load, write/read ratio and conflict rate), the Cloud-TM will accordingly release the additional nodes acquired from the cloud infrastructure and switch back to use a primary-backup scheme (that outperforms multi-master approaches in these workload conditions). This will avoid underutilizing the hired resources and minimise the operational costs sustained by the cloud user.

5. Current research directions explored in the Cloud-TM project

The design of self-tuning transactional replication mechanisms encompasses two major problems:

(1) determining the optimal replication strategy/operational mode on the basis of the current, and foreseen, workload and system characteristics
(2) allowing the efficient switching among multiple protocols/system configurations.

Concerning the first of the above two issues, we are faced with a complex multi-variable online optimisation problem that entails monitoring/tuning of a large number of system indicators/parameters.

The performance of a replicated transactional system is, in fact, not only determined by the current demand/utilisation of system resources, for example CPU, network bandwidth, memory. Workload characteristics (for example, transaction conflict probability), as well as scale (i.e. number of nodes) and topology (for example, single data center vs multiple geographically distributed clusters) of the platform have a large impact on performance.

Further, self-tuning a replicated transactional system necessarily entails the automatic optimisation of at least some of its fundamental building blocks, such as the underlying Group Communication System (Miranda et al. 2001) and local concurrency control algorithm (Sonmez et al. 2009).

We are currently tackling this issue by using differentiated methodologies, including analytical performance modelling (Sanzo et al. 2011), machine learning techniques (using both online reinforcement learning techniques and off-line approaches, such as neural networks or decision trees [Couceiro et al. 2010]), and more recently control theoretic techniques (Goebel et al. 2009).

Some relevant results that we have recently achieved in these areas are related to the performance forecasting of popular transactional memory algorithms (Sanzo et al. 2011) (for example the commit-time locking algorithm used in TL2 [Dice et al. 2006]), of Atomic Broadcast protocols (Couceiro et al. 2009; Romano and Leonetti 2011), and of alternative certification replication schemes (Couceiro et al. 2011).

Let us now analyse the second issue, namely the need to design efficient mechanisms aimed at supporting the dynamic switching among different replication algorithms. Achieving this goal requires addressing problems of both a theoretical and practical nature.

On the practical side, it requires engineering a framework that permits efficient switching between heterogeneous (possibly) distributed state consistency mechanisms. On the theoretical side, it would be desirable to design ad hoc schemes to preserve correctness despite the simultaneous coexistence of incompatible contention management strategies.

In fact, unlike in conventional non-distributed transactional systems (Guerraoui et al. 2005), the Cloud-TM platform is composed by distributed nodes, which need to consistently cooperate according to a common replica coordination scheme. Generally speaking, this raises the problem of enforcing a distributed agreement among the platform nodes in case of alteration of the adopted replication protocol.

In practice, however, the complexity of this dynamic reconfiguration task depends on the nature of the replication protocols from/towards which the system transitions. In fact, Couceiro et al. (2011) show that by exploiting knowledge of the dynamics of the destination and target replication protocols, it is possible to architect the replication manager to simultaneously support multiple atomic broadcast based replication protocols. We are currently working on mechanisms allowing for seamless switching between single and multi-master strategies, avoiding any stall or unnecessary transaction abort during transition periods.

It is worth highlighting that the efficiency of the mechanisms employed to support the system's reconfiguration is of paramount importance. This is particularly true in case the self-tuning mechanism relies on online reinforcement learning techniques for identifying the reconfiguration policies to be used in the system.

These approaches are particularly attractive as they are based on lightweight algorithms, and since they spare the burden of developing accurate models of the system to be controlled. The latter is a complex, error-prone and time consuming task, which may even be unfeasible in virtualised environments (as typical of IaaS cloud infrastructures [Amazon

2010]), where little or no knowledge is available on the actual characteristics of the underlying physical infrastructure.

On the other hand, online learning approaches require (possibly periodical) exploratory phases, in which multiple reconfiguration strategies are tested in order to build (or enhance) the initial (or current) statistical knowledge of the system's behaviour (typically in terms of rewards associated with state/action pairs [Sutton and Barto 1998]). The practical viability of these approaches is therefore strictly dependent on the availability of mechanisms supporting efficient distributed reconfigurations.

6. Conclusions

Elastic scaling, one of the key drivers underlying the success of the cloud computing paradigm, makes the problem of self-tuning the data consistency mechanisms a crucial issue in order to maximize efficiency of transactional data grids. This is one of the key goals being pursued in the Cloud-TM project, and on which this book chapter has focused.

Based on a critical analysis of the main existing transactional replication schemes, and via an experimental study relying on fully-fledged prototypes, we have highlighted a number of relevant trade-offs among the solutions proposed so far in literature. We have shed light on some of the key challenges that need to be addressed to achieve this ambitious goal, and overviewed some of the approaches that we are currently investigating to address these challenges.

Acknowledgments

This work was partially supported by FCT (INESC-ID multi-annual funding) through the PIDDAC Program funds and by the European Commission through the Cloud-TM project (FP7-257784).

References

Amazon. 2010. “Amazon elastic compute cloud amazon ec2”. http://aws.amazon.com/ec2/ (accessed October 14, 2011)

Armbrust M., Fox A., Griffith R., Joseph A. D., Katz R. H., Konwinski A., Lee G., Patterson D. A., Rabkin A., Stoica I. and Zaharia M. “Above the Clouds: A Berkeley View of Cloud Computing.” Technical Report UCB/EECS-2009-28, EECS Department, University of California, Berkeley, 2009.

Attiya, H., Gramoli, V. and Milani, A. "A provably starvation-free distributed directory protocol". In Proceedings of the 12th international conference on Stabilization, safety, and security of distributed systems (SSS'10). Springer-Verlag, Berlin, Heidelberg, 2010.

Bernstein, P. A., Hadzilacos, V. and Goodman, N. "Concurrency control and recovery in database systems." Boston: Addison-Wesley Publishing Co, 1986.

Bloom, B. H. "Space/Time Trade-Offs in Hash Coding with Allowable Errors." *Communications of the ACM* 13 (7): 422–426.

Bocchino, R., Adve, V. and Chamberlain, B. "Software transactional memory for large scale clusters". In Proceedings of the 13th ACM SIGPLAN Symposium on Principles and practice of parallel programming (PPoPP '08). ACM, New York, NY, USA, 2008.

Cachopo, J. and Rito-Silva, A. "Versioned Boxes as the Basis for Memory Transactions." *Science Computer Programming* 63(2) (2006):172–185.

Carvalho, N., Romano, P. and Rodrigues, L. "Asynchronous lease-based replication of software transactional memory". In volume 6452 of *Lecture Notes in Computer Science*, 376–396. Springer Berlin/Heidelberg, 2010.

Ciciani, B., Dias, D. and Yu, P. "Analysis of Concurrency-Coherency Control Protocols for Distributed Transaction Processing Systems with Regional Locality." *Software Engineering, IEEE Transactions on* 18 (10) (1992): 899–914.

Couceiro, M., Romano, P., Carvalho, N. and Rodrigues, L. "D2stm: Dependable distributed software transactional memory." In PRDC'09: Proceedings of the 15th Pacific Rim International Symposium on Dependable Computing, Shanghai, China, 2009.

Couceiro, M., Romano, P. and Rodrigues, L. "A machine learning approach to performance prediction of total order broadcast protocols". In SASO'10: Proceedings of the 4th IEEE International Conference on Self-Adaptive and Self-Organizing Systems, Budapest, Hungary, 2010.

—. "Polycert: Polymorphic self-optimizing replication for in-memory transactional grids". In Middleware '11: Proceedings of the ACM/IFIP/USENIX 12th Middleware Conference, Lisbon, Portugal, 2011.

Defago, X., Schiper, A., and Urban, P. "Total order broadcast and multicast algorithms: Taxonomy and survey." *ACM Computing Surveys* 36 (4) (2004): 372–421.

Dice, D., Shalev, O., and Shavit, N. "Transactional locking ii." In Distributed Computing, volume 4167 of *Lecture Notes in Computer*

Science, edited by Dolev, S., 194–208. Springer Berlin / Heidelberg. 2006.

Franklin, M. J., Carey, M. J. and Livny, M. "Transactional Client-Server Cache Consistency: Alternatives and Performance." *ACM Transactions on Database Systems* 22 (3) (1997): 315–363.

Goebel, R., Sanfelice, R. and Teel, A. "Hybrid Dynamical Systems." *Control Systems IEEE* 29 (2) (2009): 28–93.

Gray, J., Helland, P., O'Neil, P. and Shasha, D. "The dangers of replication and a solution". In Proceedings of the ACM SIGMOD international conference on Management of data, J. Widom (Ed.). ACM, New York, NY, USA, 1996.

Guerraoui, R., Herlihy, M. and Pochon, B. "Polymorphic contention management." In Distributed Computing, volume 3724 of *Lecture Notes in Computer Science*, edited by Fraigniaud, P., 303–323. Springer Berlin / Heidelberg, 2005.

Guerraoui, R. and Rodrigues, L. *Introduction to Reliable Distributed Programming*. New York: Springer, 2006.

Herlihy, M. and Sun, Y. "Distributed Transactional Memory for Metric-Space Networks." *Distributed Computing* 20 (2007): 195–208.

Kallman, R., Kimura, H., Natkins, J., Pavlo, A., Rasin, A., Zdonik, S., Jones, E., Madden, S., Stonebraker, M., Zhang, Y., Hugg, J. and Abadi, D. "H-store: a high-performance, distributed main memory transaction processing system." *Procedings of the VLDB Endowment* 1 (2) (2008): 1496–1499.

Kemme, B. and Alonso, G. "A Suite of Database Replication Protocols based on Group Communication Primitives." In Proceedings of the The 18th International Conference on Distributed Computing Systems (ICDCS '98). IEEE Computer Society, Washington, DC, USA, 1998.

—. "Don't Be Lazy, Be Consistent: Postgres-R, A New Way to Implement Database Replication." In Proceedings of the 26th Very Large Data Base Conference, Cairo, Egypt. ACM, 2000.

Miranda, H., Pinto, A. and Rodrigues, L. "Appia, a Flexible Protocol Kernel Supporting Multiple Coordinated Channels." In Procedings of ICDCS'01, Phoenix, Arizona. IEEE, 2001.

Nicola, M. and Jarke, M. "Performance Modeling of Distributed and Replicated Databases." *IEEE Transactions on Knowledge and Data Engineering* 12 (4) (2000): 645–672.

Patino-Martínez, M., Jiménez-Peris, R., Kemme, B. and Alonso, G. "Scalable Replication in Database Clusters". In Proceedings of the International Conference on Distributed Computing (DISC), 315– 329. New York: Springer, 2000.

Pedone, F., Guerraoui, R. and Schiper, A. "The Database State Machine Approach." *Distributed and Parallel Databases* 14 (1) (2003): 71–98.

Romano, P. and Leonetti, M. "Self-tuning Batching in Total Order Broadcast Protocols via Analytical Modelling and Reinforcement Learning." *ACM Performance Evaluation Review*, to be published 2011.

Sanzo, P. D., Ciciani, B., Palmieri, R., Quaglia, F. and Romano, P. "On the Analytical Modeling of Concurrency Control Algorithms for Software Transactional Memories: The Case of Commit-Time-Locking." *Elsevier Performance Evaluation Journal*, to be published 2011.

Schneider, F. B. *Replication management using the state-machine approach*. ACM Press/Addison-Wesley Publishing Co., 1993.

Shavit, N. and Touitou, D. "Software transactional memory." *Distributed Computing* 10 (1997): 99–116.

Sonmez, N., Harris, T., Cristal, A., Unsal, O. S. and Valero, M. "Taking the heat off transactions: Dynamic selection of pessimistic concurrency control". In Proceedings of the 2009 IEEE International Symposium on Parallel & Distributed Processing. Washington, DC, USA. IEEE Computer Society, 2009.

Stonebraker, M., Madden, S., Abadi, D., Harizopoulos, S., Hachem, N. and Helland, P. "The end of an architectural era: (it's time for a complete rewrite)." In Proceedings of the 33rd international conference on Very large data bases (VLDB '07). VLDB Endowment 1150–1160, 2007.

Sutton, R. S. and Barto, A. G. *Reinforcement Learning: An Introduction*. Cambridge MA and London: MIT Press, 1998.

CHAPTER THREE

DATA-INTENSIVE STORAGE SERVICES ON CLOUDS: LIMITATIONS, CHALLENGES AND ENABLERS

ELLIOT K. KOLODNER, ALEXANDRA SHULMAN-PELEG, DALIT NAOR, PER BRAND, MICHEL DAO, ALBERT ECKERT, SPYRIDON V. GOGOUVITIS, DANNY HARNIK, MICHAEL C. JAEGER, DIMOSTHENIS P. KYRIAZIS, MIRKO LORENZ, ALBERTO MESSINA, AIDAN SHRIBMAN, SIVAN TAL, ATHANASIOS S. VOULODIMOS AND YARON WOLFSTHAL

The emergence of cloud environments has made feasible the delivery of online computational services that provide access to virtualised resources. However, current approaches do not tackle key issues related to cloud storage which are of major importance given the amount of data produced by various sources (e.g., smart phones, cameras, social networks). In this chapter, we discuss the challenges of providing optimised data-intensive storage services and enablers to meet them. We are building a scalable and flexible infrastructure facilitating a new data model to raise the abstraction level of storage, and allowing data mobility, computational and content-centric access to storage, as well as mechanisms for cost-efficiency with provisions for QoS and security guarantees. Here, we describe the need for these innovations and present the technologies and architectural enablers that support them. Finally, we demonstrate how these technologies and innovations address a large spectrum of use cases and economic and business trends.

1. Introduction

The explosion of personal and organisational digital data has been recognised as one of the most significant trends of the decade. Generated data is growing faster than we can store it, while in parallel our society has become critically dependent on services requiring this data. In the emerging era of the Future Internet, the explosion of raw data and the dependence on data services is expected to be further amplified due to the increasing proliferation of data-intensive services and the digital convergence of telco, media and IT.

The proliferation of data is characterised not only by the rate of growth but also by a shift in data characteristics (Gantz 2008). Rather than being primarily transactional and structural data, today unstructured data accounts for more than 90% of the accumulated storage, and content depots are becoming the majority. Internet photo sharing sites (for example Flickr and Facebook) store billions of images and petabytes of data. YouTube accounts for 100 million daily video streams, almost as much as all medical imaging. Amazon S3 presently stores over 446 billion objects (Amazon 2011b) and is expected to store 1 trillion by 2012.

"Cloud Storage" is a model for online networked storage delivery and management, where storage is provided in a "storage-as-a-service" pay-per-use manner. The storage infrastructure is hosted remotely by a provider on multiple data centers; in the case of a public cloud, the storage provider is typically a third party. Businesswise, it is driven by lower Total Cost of Ownership (TCO) for storage, as measured in dollars/GB. The pricing models for cloud storage services are typically based on the storage capacity and bandwidth/number of accesses.

Emerging cloud-based infrastructures for storage have been widely accepted as the next-generation solution to address the proliferation of data and the reliance on it. Such infrastructures promise essentially infinite scalability and capacity, with continuous and ubiquitous availability at substantially lower costs than traditional storage solutions. Nevertheless, today's cloud environments need to overcome specific limitations. In this chapter, we start by highlighting these limitations and the corresponding challenges. We then focus on enablers to overcome them. We present an overview of an architecture addressing these enablers, which the VISION Cloud project (VISION Cloud 2011) is currently developing. Finally, the chapter concludes with the description of use cases that are demonstrated via the VISION Cloud infrastructure.

2. Shortcomings of Existing Systems

To demonstrate the limitations of today's infrastructures and to motivate the objectives of our work we present a futuristic scenario examining a day in the life of a citizen, Ann, who strongly relies on the wealth of services enabled by the Future Internet in 2013. Ann is a journalist for a large European broadcasting company that delivers news via TV, radio and web, and depends on reliable, secure and efficient access to data from anywhere at any time, regardless of the data's physical location and representation. Below is a sample of tasks that Ann has to perform together with the description of the capabilities required to achieve them that are missing in today's cloud offerings. We describe enablers to provide these capabilities in Section 3.

Task 1: Ann's company has recently acquired two new media-rich news channels aiming to improve news coverage for subscribers across Europe. The content of these channels must be integrated into the company's network and be supported by the consolidated IT infrastructure including the company's website.

New capabilities required:

- **Data mobility and Federation:** This task requires the ability to seamlessly integrate the two new channels, for example employing mechanisms that allow searching and using the content from the repositories for the two channels without the need to move, reformat or change the attributes of the data. Clouds today provide very limited mechanisms to interoperate and federate across other clouds.
- **Computation:** This task requires the ability to uniformly analyse the content from the two sources, for example automatically adding and extracting metadata to the acquired content in the background or on demand. The ability to perform computations close to the storage for the content would facilitate the execution of this task.

Task 2: As new items are posted by journalists and read by viewers, the channels analyse their popularity. While popular items have multiple replicas in the relevant geographic locations, non-popular items may be moved to storage in less expensive locations to reduce storage costs.

New capabilities required:

- **Extensible data model**: System-level usage at the granularity of an entire object must be aggregated and stored with the objects, including viewer profiles, hours etc. A flexible data model that is defined and accessed by the user of the infrastructure (rather than the infrastructure administrator) must be in place to support this aggregation.
- **Computation**: The news channels need to perform analytics over the usage statistics of the data, its capacity and popularity to smartly determine relocation strategies for the data. Again, the ability to perform computations close to the storage holding the data would facilitate the execution of these tasks.
- **Quality of Service**: For the best viewing experience, media items are served from the replicas closest to a viewer's geographic location. There is a need for storage management mechanisms that will ensure QoS by dynamically optimising the location and the number of object replicas. These mechanisms should include algorithms for automatic placement of object replicas based on multiple parameters, such as resiliency, access patterns, and geographic restrictions expressed in an SLA. In addition to ensuring QoS, they should reduce the overall storage expenditure by using less costly IT where appropriate, as defined by the SLA mechanisms (for example the requirements for streaming news data to large video devices are different from streaming to mobile devices).

Task 3: Ann's management decides to launch a campaign to publish the new integrated content offering. Ann is asked to prepare a short film to advertise the acquisition. The film will include a short historic summary of the new channels, using news items taken over their twenty years of operation.

New capabilities required:

- **Content-based access to storage:** This task requires searching for key events of interest that were captured and stored over the years in the archives of the companies. The ability to access objects, old or new, in a federated cloud, based on their content and relationships regardless of their physical location, representation and type, would enable performing this task in a simple and efficient manner. New relationships should also be integrated into existing content networks during the analysis of newly acquired data.

3. Objectives and Innovations

We have identified a few enabling technological areas that require progress and innovation beyond what is currently offered by cloud infrastructures. Below we outline each area, pointing out the gaps and proposing how the VISION Cloud project can develop them further.

3.1 Raising the abstraction level of storage

Storage has been accessed traditionally via two main types of interfaces: block and file. The low-level block interface enables basic read and write operations, and treats the storage as an unlimited array of raw bytes. The higher level file system interface treats storage as containers of semantically related bytes, with a directory structure relationship among the containers. Modern file systems also introduce extended attributes to files as a mechanism to associate a limited amount of additional metadata with a file. These two data models have been optimised for scale and performance.

Over the last decade, a new model of object storage has been introduced for access to storage devices (SNIA 2009). This model has been successfully adopted for the cloud, replacing the traditional file system and adapting it to cloud scale (e.g. Amazon 2011a; OpenStack 2011; EMC 2011). It flattens the tree hierarchy, which is no longer relevant, and relaxes semantics and consistency. This new model considerably simplifies access to storage, as it does not require a special client, file manipulation, or mounting of a file system. The models that have emerged so far are not very rich with respect to metadata and usually limit its size, attributes and query options. Furthermore, the security model of most of these systems is weak and does address the issue of isolation among multiple tenants hosted on the same infrastructure.

With the fast growth of the number of objects stored by each user, the efficient management and attribute-based retrieval for the objects becomes more and more acute. For example, S3 currently holds billions of objects, which divided by the number of US residents means 1,440 objects per user (Amazon 2011b). How can users efficiently look-up an object based on its properties? How can they purge and delete old objects, or sort them by importance and value? How can they securely share them with others? Although many of these issues are still open problems, some solutions have evolved to address this management challenge. For example, Amazon allows the deployment of an additional layer consisting of a SimpleDB database (Amazon 2009) providing the query functionality of a database over the S3 storage service. However, this requires an additional application layer to programmatically connect between the storage service and the database service, rather than being an integral part of the storage service offered to the end client. Other existing solutions are either implemented above the storage as an SaaS layer or mount the cloud storage back to the user as a local disk and file system (Jungle Disk 2011; DropBox 2011; Gladinet 2011).

We are working on a more powerful data model that fits the scale of the cloud yet has a rich set of metadata and access methods, and supports a strong security model. This data model is optimised for immutable data, maintains a global namespace for the objects, and supports object versioning. Furthermore, it associates system and user metadata (in the form of [key, value] pairs) with objects, and leverages a NoSQL table service incorporated in the storage service to provide access to objects through this metadata. For example, in addition to the basic "put" and "get" operations, it supports List-By (key), and List-By (key, value-range) over collections of objects. Future extensions could include the association of a schema with the user metadata. The metadata allows for highlighting of the content so that it can better fit with the application (the cloud managing the objects is no longer agnostic to the data type, such as the case for file systems and object stores). Furthermore, the proposed data model offloads some capabilities traditionally provided by the middleware layer, down to the storage layer. The motivation for this is twofold. First, from the point of view of the middleware, there is no longer a need to federate over the namespaces of multiple storage repositories or to maintain the association between metadata and data in databases external to the storage. Second, from the point of view of the storage, it enables optimisations not previously attainable such as collocating data and metadata and minimising data loss by keeping the metadata and data under the responsibility of one system. Finally, the data model supports a distributed security model capable of delegation and federated identity, which are key requirements currently missing from existing cloud infrastructures. The data model API is in the spirit of the newly emerging storage cloud standard of CDMI (SNIA 2010).

In summary, technology today allows for the easy storage of virtually infinite capacities of personal and business data for reasonable costs; however, no adequate mechanisms exist for efficiently managing the data. Can this data be easily accessed after some time? Can this data ever be deleted? Raising the level of storage abstraction enables the incorporation of built-in mechanisms required to efficiently manage the infinite capacity of data in the cloud.

3.2 Data mobility and federation without boundaries

In order for virtualisation platforms and cloud infrastructures in particular to fulfill their promise, they must allow providers and users to rapidly redeploy and move resources. While this is partially achievable for compute resources today (such as VMs), it is not the case for storage and data. In the absence of true data mobility, users cannot easily migrate their data across providers and thus suffer from data lock-in (Armbrust et al 2009), one of the most significant obstacles hindering wider adoption of cloud services. The capability of data mobility is also fundamental to addressing IT evolution and heterogeneity, as data needs to migrate between different platforms. Furthermore, this capability is also a key for federation and interoperability between providers and systems.

Barriers to data mobility are both technological and business-related. On the technology side, moving data is slow and prone to failure when large amounts are involved. In addition, data model and API inconsistencies prohibit interoperability. On the business side, there is usually no economic incentive for the data provider to allow easy migration of user data. New technologies to address these barriers are needed, including efficient methods for large data migration as well as cloud federation and interoperability.

3.2.1 Efficient methods for large data migration

Moving large volumes of data (in the order of terabytes) between remote locations over wide area networks, while ensuring QoS in the presence of unpredicted network latencies, is not reliably feasible today. In VISION Cloud we are overcoming these limitations by employing new mechanisms such as network de-duplication as well as extending lazy copying techniques and network overlays, and leveraging the information captured by metadata for executing data transfers intelligently. A key challenge in this approach is how to exploit these data reduction techniques while preserving privacy (Hamik et al. 2010). As part of our research in VISION Cloud, we have identified attacks that exploit client-side deduplication and propose mechanisms to prevent them (Halevi et al. 2011). For example, we show how deduplication can be used as a side channel that reveals information about the contents of files of other users or serves as a covert channel by which malicious software can communicate with its control centre, regardless of firewall settings on the attacked machine (Hamik et al. 2010). Another, even more severe data leakage problem occurs when an attacker who knows the hash signature of a file can convince the storage service that it owns that file, hence the server lets the attacker download the entire file. To overcome this attack, we introduce the notion of proofs-of-ownership (PoWs), which allows a client to efficiently prove to a server that it actually holds a file, and not just the hash signature (Halevi et al. 2011).

3.2.2 Cloud Federation and Interoperability

Creating a federation of existing clouds allows the provision of a unified storage and data service over the clouds. This is important for various use cases, including mergers and acquisition, consulting and collaboration. Federating the resources of cloud providers may also help to scale beyond the limits of a single cloud, providing higher reliability, availability and security. Collaboration agreements that allow one cloud provider to borrow the resources from another can help meet SLAs and also give customers the appearance of infinite capacity, while avoiding over-provisioning and remaining cost effective. So far, this approach has been studied primarily for compute clouds (rather than storage). For example, several options for cloud federation architectures have been proposed, such as resource federation (Rochwerger et al. 2009 and 2011), where the clouds use each other internally or a cloud broker topology, where there is an intermediate layer that federates separate "back end" clouds (Elmroth and Larsson 2009; Buyya et al. 2010; Elmroth and Tordsson 2005). Although some work addresses these issues in the context of the GRID or Cloud VM, the federation of storage resources is still in its infancy. Another level of federation is at data access, where users can access their resources through any cloud in the federation. One recent example of such a system is RACS, which is a proxy that transparently spreads the storage load over many providers (Abu Libdeh et al. 2010). It stripes user data across multiple clouds by applying RAID-like techniques at the cloud storage level. Other recent examples are jclouds (jclouds 2011) and Libcloud (Libcloud 2011), which are open source libraries implementing common cloud APIs.

Recently, the emerging need for hybrid clouds that combine public and private clouds has also been recognised (SNIA 2010; Silk 2010). These clouds essentially federate multiple pools of resources in order to meet business and performance requirements. For example, sensitive data or data with low latency requirements is stored in a private cloud, while less sensitive data or data that has less strict latency requirements is stored in a public cloud, thereby exploiting its low cost and elasticity characteristics. Similarly, replicating data in a private cloud to an external public cloud may lower the overall cost of data backup and archiving. In addition, hybrid storage clouds are useful when elastic demand for storage exceeds the available internal resources (SNIA 2010).

3.3 Computational storage

Compute and storage are in most cases treated as two different resources in a decoupled manner. Given that bandwidth is neither infinite nor networking costs negligible, it is clear that for many applications it is better to move the computation to the data, rather than bring the data to the computation. As the cloud model has emerged, this idea of bringing compute to storage has been applied for restricted programming paradigms, e.g. MapReduce (Dean and Ghemawat 2004) or specifically for key-value storage services, e.g. Comet (Geambasu et al. 2010). In addition, approaches such as Isard et al. (2007) study how to utilise resources in a large cluster by executing data-parallel programs. We propose a more general paradigm.

The primary role of computation in emerging data-intensive storage environments is to serve the data by analysing, refining, and transforming it, discovering correlations and relations, and then associating this knowledge with the data through metadata annotations and derived summarising semantic data structures. Accordingly, VISION Cloud introduces a model for computational storage, based on the following main principles: (a) computations are executed close to their data, since the size of an encapsulated computation is typically small compared to the size of the data it is accessing (ideally computation is co-located so no networking resources are consumed); (b) high utilisation of cloud resources enabling parallelism where appropriate; (c) high-level control by the users allowing control of computation through policies without the need to start, stop or manage individual computations. For example, computations can be injected into the cloud and instructed to analyse all data objects of a given type or within a given context (as characterised within the data model).

These principles are supported through a programming model for computational agents that run in the storage, called "storlets," originally introduced in the context of digital preservation repositories (Rabinovici Cohen et al. 2008). A storlet is released into the cloud and activated by events on data. A storlet defines not only a computation but also triggering conditions whereby it is activated (for example when accessing or updating a given object, the creation of a data object with a given metadata tag, the addition of a given metadata tag to an existing object, etc.), constraints that apply during its lifetime (for example maximum CPU usage), its input and output data objects as well as the necessary credentials to access them. Finally there is a management interface for storlets to deal with aspects such as billing and accounting.

A storlet is potentially very long-lived and may be repeatedly activated as defined by its triggering conditions, perform some computation, and then become passive again. A runtime environment schedules and executes a storlet securely in a sandbox that enforces constraints and mediates between the storlet and other platform services, for example to create and access an object and update its metadata. Regarding the model for fault-tolerance, when a storlet is passive it is treated as a data object (i.e. replicated) and when active it is monitored and upon failure is re-executed from the checkpointed state when it was last passive.

The storlet programming model subsumes traditional batch-job computations through the special case in which a storlet's triggering condition is already met at the time it is inserted into the cloud. In the design, storlets can also (when triggered) create massive parallel computations in the manner of MapReduce, Dryad or PACT/Stratrosphere (Isard et al. 2007)—though only MapReduce is planned in the scope of the VISION Cloud project.

In summary, VISION Cloud is improving the functionality and performance of data intensive services by providing storlets, a means to bring computation close to its data. A runtime environment ensures the secure execution of storlets, triggers storlets at the appropriate time, enforces isolation constraints, and supports application and cloud management.

3.4 Content-Centric Access

Alongside the higher abstraction level of storage, VISION Cloud provides cloud-scale content-centric access to data objects. This innovative concept addresses the observation that in a highly virtualised infrastructure, requests for data (from users or applications) should ideally be unaware of the details of the data's container, for example location (fileserver name and/or address, file, URL), format, storage representation, and the implementation details of the low level protocols that transfer data. Rather, in order to best leverage the virtual nature of the cloud, service requests should access data by abstract content (and domain) related names, whether this is the title of a book or a research paper, or the identification of a patient for a medical record. This approach is similar to content-centric networking as proposed by Van Jacobsen et al. (2009). In order to cope with the future growth of the Internet, content-centric networking proposes that the user request data by an identification of its content rather than its location as done currently with URLs. The infrastructure of the network is aware of the data location and can optimise placement according to content demands. Analogous to the networking approach, the storage infrastructure of VISION Cloud is fully aware of the physical details. A unit of content may be composed of many files, may be replicated, possibly across different locations and file systems, and may be accessed by different types of protocols.

Thus, the VISION Cloud infrastructure enables efficient retrieval of objects based on their content, properties and the relationships among them. Virtualisation is essential for providing this capability. Specifically, as the objects' content, properties and interdependencies change, their storage location, representation and access path change as well; however, VISION Cloud virtualises the storage to ensure that these implied changes are transparent to the application and have minimal performance implications. Changes in the underlying physical storage network, for example replication or migration of objects, do not affect content-centric access to them and is carried out in a completely orthogonal and transparent manner from the perspective of users and applications. Lastly, VISION Cloud provides the ability to build and maintain content networks, consisting of sets of data objects, their associated metadata and the relationships between them.

While highly desirable, content-centric access is a very challenging and ambitious property to achieve at cloud-scale infrastructure level. Companies have made some initial attempts at such systems already. For example, YouTube allows for content-centric access to video content; instead of providing the user with a hierarchy of categories for video content, the user can search for keywords, previews and ratings about the content. However, as seen with the YouTube example, such approaches are largely proprietary and domain specific. VISION Cloud goes one step further and provides generic mechanisms by which infrastructure developers and service providers can guide the infrastructure management system to apply storage optimisations for specific content domains, leading to simplified and efficient storage access.

Also, in research there has been an initial attempt at content centric access to storage. In particular, the CIMPLE system (Delaet and Joosen 2009) proposes a content centric storage interface for a file system in place of the traditional file system interface. In contrast, VISION Cloud targets content centric access to a data store in large-scale deployments by using cloud computing technology.

Altogether, VISION Cloud aims to develop powerful content-based access technology, with the following strengths:

- Scalability—addressing Internet-scale systems with Internet-scale sets of objects, containers, relationships and properties, without sacrificing performance by the use of virtualisation and content-centric access.
- Interoperability—supporting the federation and fusion of multiple disparate content networks that are not bound to location-based addressing schemes.
- Optimisation ready—ability to define specialised content domains (for example GIS-related geo-coordinate handling, healthcare- or media domain based addressing) which drive domain-specific storage optimisations.
- Dynamic—capable of dynamically adding metadata on-the-fly to reflect changing content and container properties without sacrificing access performance and scalability.

3.5 Capabilities for Cloud-based Storage

Although cloud storage available from providers such as Amazon and Google offers useful features such as demand-based access to raw storage resources, it is not ready to store the critical data of individuals, businesses and governments with the required reliability and QoS as evidenced, for example, by the rudimentary SLA of Amazon S3 (Amazon 2011a and 2011c). We are developing technologies necessary to close the above gaps, addressing aspects of QoS and security assurance as required for business critical and sensitive applications while building a cloud infrastructure that continues to maintain the cloud spirit (e.g. virtualisation, pay-per-use, scalability, etc.).

Current cloud storage providers offer SLAs that guarantee service availability and give service credit or refunds for lack thereof, but do not address data availability and protection (Nirvanix 2011; Amazon 2011c; Microsoft 2010). Research shows that some approaches have been proposed for cloud storage SLA schemes and frameworks. For example, in one work various SLA parameters are described for four types of services that can be offered in cloud computing environments (Alhamad et al. 2010a), and in another work these are taken into consideration in the architecture and protocol proposed for an SLA-based trust model for cloud computing (Alhamad et al. 2010b). Interesting approaches to managing the mapping of low-level resource metrics to high-level SLAs have been proposed (Brandic et al. 2010; Emeakaroha et al. 2010), where the frameworks are embedded into FoSII infrastructure, which facilitates autonomic SLA management and enforcement. However, the majority of this state-of-the-art is mostly research work and not supported by the current major commercial offerings.

VISION Cloud requires an SLA framework that takes the stored content and content-centric access into account. Thus, the SLA framework uses terms beyond those currently typical to SLAs, such as cost, availability, and performance, and also includes terms regarding content-related information (for example different content types may be linked in an SLA with different pricing models and performance estimations for data transfer). Furthermore, for VISION Cloud we are developing mechanisms for proactive SLA violation prevention. To this end, a monitoring specification will define the information that needs to be monitored. The specification will be derived automatically and dynamically based on models and patterns regarding the SLA terms. The monitoring system itself is also advanced beyond being a simple messaging system to aggregating, applying rules and extracting valuable information from the analysis, and to being modular, i.e. allowing new sources of information to be included as the need arises.

Other architectural challenges include: supporting multi-tenancy, where a massive number of users share the same storage infrastructure, guaranteeing secure and authorised access to the data and services, and providing tools for checking compliance with standards and regulations. Most current work concerning compliance in the cloud, for example Cloud Security Alliance (CSA 2011), reflects the effort to apply the common IT Security management practices (which are driven from standards and specifications such as SAS-70 Type I and Type II Attestation, ISO 27001 Certification, PCI DSS Compliance) to the cloud paradigm. However, that process is quite hard and has not yet led to universally recognised practices and methodologies. VISION Cloud improves the state-of-the-art for compliance in the cloud by applying the innovative concept of SIEM (Security Information and Event Management) specifically to the cloud storage infrastructure, introducing as much automation as possible in the processing of audit events and in compliance reporting. In particular, XDAS (Open Group 1998), an existing standard and interoperable format for log data, specific for collecting audit and regulatory information (events), and open source technologies enabling log collection and processing are adopted.

In addition, special management tools are needed to optimise the storage services and provide them at low cost, for example automatic placement of data, managing energy efficiency, and automation, provisioning and operation of the underlying storage. Finally, the architecture must provide hooks for accounting and billing of the storage services for measuring, charging and reasoning about their cost.

4. Architecture and Prototype Implementation

Below we describe the high-level architecture and some implementation details that allow the ideas and enablers presented in Sections 2 and 3 to be realised. The full details of the architecture can be found elsewhere (Kolodner et al. 2011).

We conceptualise the architecture in two dimensions. In the first, the architecture has a logical separation between data-related operations (e.g. adding, changing and deleting data) on one hand and management operations (e.g. service provisioning and monitoring) on the other. In the second dimension, each of these has an external access/interface layer that provides access to users and applications, and an internal operating layer that serves these external requests as well as executes autonomous operations such as dynamic optimisations, load balancing and monitoring.

Fig. 3.1 presents a high-level view of the architecture including the interfaces towards applications, users and administrators, the layers of the architecture and the interactions between the layers. This architecture introduces two complementary services, namely the Data Service and the Management Service, which together provide the functions of a cloud for data-intensive services. The Data Service, including the Data Access Layer (DAL) and the Data Operating Layer (DOL), enables manipulation of data objects and their properties, computation on storage, mobility, availability, reliability and security. The DAL provides unified interface to data across the clusters of a cloud, and encapsulates the DOL which realises the data service over a set of distributed heterogeneous physical resources.

Fig. 3.1. High Level Components Overview

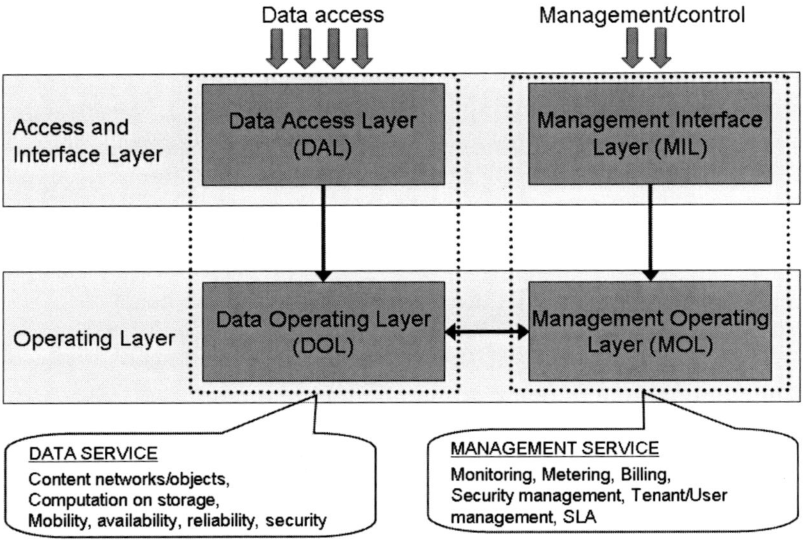

The Management Service, including the Management Interface Layer (MIL) and the Management Operating Layer (MOL), enables service provisioning and monitoring, accounting and billing, security management and transformation of user-specified service level requirements to management operations on the underlying infrastructure level. This service is also distributed across the clusters of a cloud. The MIL, just like the DAL, provides a unified interface to management of services across the clusters of a cloud, and encapsulates the MOL which realises the management service over a set of distributed heterogeneous physical resources. The MIL deploys management models that translate business level objectives into operating level settings and tasks.

The architectural separation between the data and management services is inspired by the unique service model of the storage cloud. In compute clouds, the management service is used to provision and manage compute resources, which interact with external entities as defined by the service provider for a particular computing task. The storage cloud is different—once the storage resources are provisioned, they may be used by different, independent service consumers through cloud storage APIs, with different characteristics and requirements on latency, throughput, availability, reliability, consistency, etc. The data service and the management service are designed to be separate and independent in order to facilitate this differentiation and provide the flexibility required to enable the innovations mentioned earlier.

The realisation of the layers is distributed. For example, an object may be stored in one cluster and accessed through a request addressed to a second cluster. The DAL in the second cluster finds the target cluster on which a copy of the object resides, and transparently redirects the request to it. Furthermore, the implementation of the layers is highly distributed and parallel; the same software stack runs on every server of a cluster, and many client operations are carried out in parallel on each server.

4.1 Physical Infrastructure

VISION Cloud is built over a set of physical resources. At the highest level, a VISION Cloud consists of multiple data centers where each may have one or more storage clusters, each of which contains physical compute and storage resources. The data centers are connected by a dedicated network. A storage cluster is composed of storage rich nodes constructed from commodity and not necessarily highly reliable hardware and connected by commodity interconnect. Reliability is ensured via the software.

4.2 Additional High Level Concepts

The data model is based on data objects that contain data of arbitrary type and size, having unique names that can be used to access them. Data objects are grouped in containers, which are used for data management purposes (to group related objects), isolation, and internal management (containers are the unit of placement; this reduces the frequency of global placement decisions and the size of location information required for routing). Metadata can be associated with both objects and containers.

Another high-level concept is the account model of VISION Cloud. The account model consists of tenants and users. A tenant is an organisation that subscribes to storage cloud services, and may represent a commercial firm, a governmental or other organisation, consisting of one or more individual persons. A user is the entity that actually uses storage services, and may refer to a person or to an application. A user belongs to one and only one tenant, although a person might own a user account in more than one tenant. A tenant administrator creates users and manages them. A user has an identifier (unique within its tenant) and may have credentials allowing him to authenticate himself to the cloud; finally, a user may create containers and data objects in them.

Wide area distribution, scalability and resiliency also play a central role in VISION Cloud. Objects are replicated across clusters and data centers to ensure their resiliency and availability despite failures, e.g of disks, servers, clusters and network links. Objects are fixed content that are written all at once and not updated. Leveraging this fact, we use a multi-master (aka symmetric) replication mechanism, where an operation can be carried out at any object replica, and its results are replicated to the other copies. Given the large-scale distribution in VISION Cloud, the consistency level provided is eventual consistency (Gilbert and Lynch 2002), which means that eventually all of the object replicas will agree on the same value for an object.

4.3 Implementing the Enablers

The architecture, design and implementation of VISION Cloud supports a system with hundreds of storage clusters, where each storage cluster can have several hundred nodes and are spread out over dozens of data centers. A local catalogue in each storage cluster keeps track of the metadata and the locations of objects in that cluster. In our initial implementation GPFS-SNC (Schmuck and Haskin 2002; Marks 2010), a distributed file system for clusters composed of servers with direct attached storage, holds the object data. A global catalogue has representatives in each cluster and keeps track of the locations of container-replicas.

Additional technologies and architectural enablers that are introduced in order support the innovations described in the previous section are now described below.

Raising the abstraction level of storage

This innovation is implemented primarily by the DOL via a catalogue, a cluster-wide service that stores the metadata for each object residing in the cluster. It provides an efficient mapping from object name to the storage location in the cluster, and manages the metadata associated with an object. It also provides list and query operations, such as listing the objects in a container according to certain rules or querying by key name and/or value. Since the catalog needs to be highly scalable, optimising performance for billions of objects, it is implemented based on Cassandra, a distributed column-oriented NoSQL database (Cassandra 2009).

Data mobility and federation without boundaries

VISION Cloud employs efficient object-level deduplication, which is enabled through the catalogue service that allows mapping of objects with the same content to the same physical storage entry. The DAL and the DOL level components can support the Proof-of-Ownership (PoW) scheme developed to prevent data leakage through deduplication (Halevi et al. 2011). The DAL supports federation and in addition, the access control architecture of VISION takes into account the access federation requirements detailed in Section 3.2.

Computational storage

Computational storage is implemented through two components: (1) the Storlet Programming Environment (SPE) that is located at DAL and (2) Storlet Execution Environment (SEE) that is part of DOL. The SPE provides users the means to define storlets and ready them for injection into the VISION Cloud where they reside. The SEE provides the underlying framework to allow execution and handling of storlets with the main focus on storlet lifecycle management operations such as storing, moving, triggering/activating and then executing. Finally, the MOL layer contains a resource allocation component, which decides where objects should be located and where storlets will be executed, taking into account optimisation of resources and subject to restrictions.

Content-centric access

Content-centric storage depends on the association of metadata with objects along with query capabilities. It also implements relations on objects, using object metadata to define the relations. Depending on the request, the content-centric component may translate a single client request to multiple internal requests, which are combined and processed before returning the response to the user.

Capabilities for cloud-based storage

Some of the advanced capabilities are implemented through the MIL and MOL, which provide functionalities such as accounting and billing and SLA Management, allowing customers to edit and negotiate SLAs, thereby posing requirements on the infrastructure. The SLA schemas can depend on content-based terms, thus further supporting the content-centric approach of VISION Cloud. Additional capabilities, such as security assurance and multitenancy are provided through authentication and authorisation components which are responsible for enforcing access control and securely isolating the requests of the various tenants throughout the processing of requests.

5. Demonstration via Use Cases

The use cases described in this section are used to validate the concepts and the innovations proposed in the VISION Cloud project. The use cases address a large spectrum of domains and business needs, emphasising converged IT/Media/Telco services.

5.1 Telco Use Case

Information and communication technology and associated industries are going through a major transformation. The focus of the telecom business has been rapidly shifting from network infrastructure to service and content management. This shift poses new demanding situations for data management for the telecom business:

- The amount of digital data and its variety that customers wish to store and manage is heavily increasing. Presently, the data consists mainly of multimedia content (e.g. pictures, music, videos), but new types of data including personal documents (e.g. bills, pay statements, scanned administrative documents) will also become important in the future. Enterprises will also have to manage different types of data particularly through backup or archiving services.
- There is a growing need to easily provide added value services to users for their digital data.

For telecommunication operators to offer data-intensive applications with high quality of service, they need to address the following technical challenges:

- Workload increase: potentially, these applications will have a number of users growing at a very fast pace as the customers adopt them (up to several million). Therefore, the underlying data storage layer should be scalable enough to cope with a rapidly increasing workload.
- Mobility and access device diversity: with the advent of more and more powerful mobile devices (e.g. smartphones, tablets, netbooks)

that are connected to telecommunication networks, customers will be able to access the data intensive applications from anywhere and from different devices depending on their current location.

- Service and data interoperability: customers will have the ability to use several services to store their content and will want to share it with friends and relatives, possibly by integrating their content into their web ecosystem. This implies that the content management applications should be able to federate content from multiple sources, to export content with security and access rights policies, to comply with regulations (local and global) and to offer rich content management capabilities.

The telco scenario demonstrates how multimedia content management services can be built by a telecommunication operator via the use of the VISION Cloud platform. It will be based on mashup services by which customers can aggregate and share their multimedia content, and on the management of software elements uploaded to the cloud storage as "storlets."

The content manipulated by the service is stored in the VISION Cloud data model and managed through the content-centric layer and by storlets that can be triggered on specific access operations and run near the content. The use case allows a customer to upload different types of content to VISION Cloud, triggering processing through storlets to transform and organise the content. From a service provider perspective, the use case will show how new services can be easily added, while taking advantage of the organisation of content within the VISION Cloud.

5.2 Media Use Case

The Media industry worldwide is in the midst of complicated convergence processes, from analogue to digital, from old business models to new markets. The need to capture, store and distribute data is at the centre of this transformation, and online video is the biggest driver. Higher storage needs are expected in the next five years that will exceed what media companies can host as single companies, thus driving migration to cloud computing and distributed storage. An inevitable transition is also expected from traditional tape-based production paradigms to entirely file-based workflows. Below we list the limitations of today's IT solutions towards satisfying these requirements.

- Current cloud-based models may fail when adopted for massive content rate applications, both in terms of the number of objects produced and distributed and in terms of their size.
- To support data visualisation, there is a need for flexibility in data formats and reachable platforms.
- There is a need for multiple interoperable content formats in media production.
- There is a need for flexibility to quickly deploy small solutions to handle published data.
- There is a need for flexibility in content management in contribution networks.
- Since data collections might come from unexpected sources, the cloud platform should be adaptable or have a mechanism to ingest and transform varied data formats into a readable format that can be mixed with other data.
- Established content indexing practices indicate a need for the introduction of baseline content management features in the storage layer.
- Support for heterogeneity is key, due to the necessity of migrating old formats to the digital domain, and allowing for the utilisation of older material in new production.

The Media scenario demonstrates ubiquitous access to existing content and ubiquitous generation and storage of new content. This is achieved through methods that allow users to search and retrieve content regardless of their position, device capability and language, enabled both by content centric storage and by computational storage. Moreover, storlets are used for on-demand adaptation of content, i.e. methods that are able to adapt content format and resolution to the various communication channels and devices. Automated content analysis and indexing is achieved through the rich data model implemented in VISION Cloud. Finally, also through its rich data model, the VISION Cloud infrastructure would act as a unified interface for all the media production and distribution operations performed by the upper-level service providers: broadcasters, media producers, and publishers alike.

5.3 Healthcare Use Case

IT for the healthcare domain is driven by the need to provide efficient, patient-centred, outcome-focused healthcare. To achieve this, IT must focus on clinical applications, cloud-based delivery technologies and innovative approaches to reducing the costs associated with IT infrastructure. Important themes identified by a recent study are the disruptive impact of (U.S.) healthcare reform, cloud computing, and storage solutions (Hanover and Dunbrack 2011). Responsibility for the nearly exponential growth of storage needs can be assigned to advances in diagnostic technologies that produce vast amounts of data, mainly digital images and electronic health records. Healthcare provider organisations will increasingly be forced to consider cross-organisational interoperability and patient driven healthcare requires that patient health records be available. In particular, patient health records are seen as a means to leverage cross-organisational interoperability, because ultimately it is the patient's health and wellness that is the product of healthcare industry. Achieving a patient's consent for the use of healthcare data for medical research and for web based add-on services could be a major success factor for specific advances in medical science and also products and services offered by healthcare related industry.

The focus of the VISION Cloud healthcare use case is therefore to evaluate how VISION Cloud's innovations will foster "personalized healthcare applications" based on patient health records. These applications provide a patient-centric view of all healthcare activities, related data and services. A patient's healthcare data is stored and used either by the patient or by any healthcare provider involved in the patient's healthcare process. In addition, third party applications may access a patient's data to provide services to all actors in the scenario. Healthcare data are commonly stored according to complex standards such as HL7, DICOM, or OpenEHR. These standards provide exact descriptions of medical information while at the same time allowing healthcare systems and medical devices to interoperate reliably and precisely.

Based on the above, the personalised healthcare scenario poses specific requirements to storage cloud environments that are met through the VISION Cloud enablers:

- The ability to store all patient-related data as native and immutable objects and make these objects and the relationships between them accessible as a data service. Here content-centric access supports relationships and access to the objects based on their relationships.

Storlets provide the means to do transformations to required data formats, and extract and associate appropriate metadata.

- The ability to hide much of the complexity of healthcare data from developers and support them to supply the data as a data service, while securing healthcare data and providing an utmost level of privacy for the patient. Here the metadata and content-centric capabilities ease the task of organising and retrieving medical data. Furthermore, storlets enable healthcare applications to efficiently perform complex transformations, aggregations and analysis on masses of healthcare data without any need to download the data to an application's local storage. In addition to reducing bandwidth requirements, this has the additional benefit that it makes it easier to protect data against unauthorised use.
- The ability to provide data to users according to their roles and context (e.g. doctors, patients, online medical service providers, researchers, pharmaceutical companies) will be provided by the advanced security policy-based mechanisms for access control. Furthermore, storlets can adapt the view of the data based on the user's role, for example a storlet can anonymise data before providing access to researchers.

5.4 Enterprise Software

There are two significant IT trends in the enterprise software that can become a differentiating factor in the market. The first is the ability to provide sophisticated analytical capabilities in real time on up-to-date fresh data. The availability of servers with very large main memories, along with the advent of Main-Memory Database System (MMDBs), now put these real time capabilities within reach. The second trend is the ability to leverage cloud computing as a means for lowering Total Cost of Ownership (TCO) of various types of business applications that are provided on-demand. One major inhibitor to progress on these two objectives is the lack of true support for multitenancy in today's cloud infrastructures, which is key to handling the business data of enterprise customers.

In this use case we demonstrate the capability to perform Multitenant Analytics On-Demand. Such a scenario is based on: (1) inherent multitenancy since it is a SaaS model; (2) data ingress and periodic refresh flow for keeping tenant data up-to-date; (3) analytics on the data in a multitenant secure setup; and (4) high availability of the data. Below we list the elements are missing from existing solutions.

- Existing solutions require that the OLAP (Online Analytical Processing) service be on the same premise with the OLTP (Online Transaction Processing) systems.
- Existing systems do not offer true multi-tenancy, guaranteeing isolation among instances and customers, together with per-user billing and accounting.
- Existing solutions do not offer varying levels of data replication including cross-site replication.

The scenario of Business Analytics On-Demand (BIOD) will be demonstrated on the VISION Cloud platform for multiple tenants. The analytics engine is based on a MMDB system running close to VISION Cloud, but outside of it. The MMDB employs VISION Cloud storage for its backing store, where it stores its massive data sets as objects. There is also periodic upload to VISION Cloud of batched transactions from each tenant's on-premise location. Storlets are employed on these batches for ETL (Extract, Load and Transform) prior to feeding the data to the BIOD MMDB. Multitenancy support ensures data separation between tenants and provides per-tenant billing and accounting for storage use. Federation and data mobility prevent vendor lock-in and allow choice of VISION Cloud providers for the BIOD system.

6. Summary and Conclusions

Personal and organisational digital data is growing at unprecedented rates. To meet the growing amounts of data and the demand for new data intensive services new approaches are required. In this chapter, we have highlighted the limitations of current approaches through a sample scenario, and have described in detail new capabilities required to meet these challenges. The VISION Cloud project is building a scalable and flexible infrastructure demonstrating these capabilities: a new data model to raise the abstraction level of storage, data mobility and federation, computational and content-centric access to storage, and advanced capabilities for cloud-based storage. On this infrastructure, we are demonstrating new emerging use cases from telecom, media, healthcare and business enterprise services that are not supported today.

Acknowledgments

The research leading to these results is partially supported by the European Community's Seventh Framework Programme (FP7/2001–2013) under grant agreement no. 257019—VISION Cloud Project.

References

Abu-Libdeh, H., Princehouse, L. and Weatherspoon, H. "RACS: A Case for Cloud Storage Diversity." Proceedings of the First ACM Symposium on Cloud Computing (SOCC), June 2010.

Alhamad, M., Dillon, T. and Chang, E. "Conceptual SLA Framework for Cloud Computing." 4th IEEE International Conference on Digital Ecosystems and Technologies (DEST) 2010a.

—. "SLA-based Trust Model for Cloud Computing", 13th International Conference on Network-Based Information Systems (NBiS) 2010b.

Amazon. Amazon SimpleDB Developer Guide (API Version April 15, 2009), Amazon Web Services Documentation, 2009. http://docs.amazonwebservices.com/AmazonSimpleDB/latest/ Developer Guide/index.html (accessed September 8, 2011).

—. Amazon Simple Storage Service (Amazon S3), Amazon Web Services, 2011a, http://aws.amazon.com/s3 (accessed September 8, 2011).

—. Amazon S3 - More Than 449 Billion Objects, Amazon Web Services Blog, 2011b. http://aws.typepad.com/aws/2011/07/amazon-s3-more-than-449-billion-objects.html (accessed September 8, 2011).

—. Amazon S3 Service Level Agreement, Amazon Web Services, 2011c. http://aws.amazon.com/s3-sla/ (accessed September 8, 2011).

Armbrust, M., Fox, A., Griffith, D., Joseph, A., Katz, R., Konwinski, A., Lee, G., Patterson, D., Rabkin, A., Stoica, I. and Zaharia, M. "Above the Clouds: A Berkeley View of Cloud Computing." UCB/EECS-2009-28, EECS Department, University of California, Berkeley, February 2009, http://www.eecs.berkeley.edu/Pubs/TechRpts/2009/EECS-2009-28.pdf (accessed September 8, 2011).

Brandic, I., Emeakaroha, V.C., Maurer, M., Dustdar, S., Acs, S., Kertesz, A., and Kecskemeti, G. "LAYSI: A Layered Approach for SLA-violation Propagation in Self-manageable Cloud Infrastructures." 34th Annual IEEE Computer Software and Applications Conference Workshops (COMPSACW) 2010.

Buyya, T.R., Ranjan, R., and Calheiros, R.N. "InterCloud: Utility-Oriented Federation of Cloud Computing Environments for Scaling of

Application Services." International Conference on Algorithms and Architectures for Parallel Processing (ICA3PP) 2010.

Cassandra. The Apache Cassandra Project, The Apache Software Foundation, 2009. http://cassandra.apache.org/ (accessed September 8, 2011).

CSA. CSA Cloud Security Alliance, 2011. https://cloudsecurityalliance. Org (accessed September 8, 2011).

Dean, J. and Ghemawat, S. "MapReduce: Simplified Data Processing on Large Clusters" 6th Symposium on Operating Systems Design and Implementation (OSDI '04), 2004.

Delaet, T. and Joosen, W. "Managing Your Content with CIMPLE—a Content-centric Storage Interface." IEEE 34th Conference on Local Computer Networks (LCN), Zurich, Switzerland, 2009.

Dropbox. 2011. http://www.dropbox.com (accessed September 8, 2011).

Elmroth, E. and Larsson, L. "Interfaces for Placement, Migration, and Monitoring of Virtual Machines in Federated Clouds." Proceedings of the 2009 Eighth International Conference on Grid and Cooperative Computing (GCC '09), IEEE Computer Society, Washington DC, 2009.

Elmroth, E. and Tordsson, J. An Interoperable, Standards-Based Grid Resource Broker and Job Submission Service, E-SCIENCE '05: Proceedings of the First International Conference on e-Science and Grid Computing, Dec. 2005.

EMC. EMC Atmos Managing Big Data in the Cloud, EMC Corporation, 2011. http://www.emc.com/storage/atmos/atmos.htm (accessed September 8, 2011).

Emeakaroha, V.C., Brandic, I., Maurer, M. and Dustdar, S. "Low Level Metrics to High Level SLAs - LoM2HIS Framework: Bridging the Gap Between Monitored Metrics and SLA Parameters in Cloud Environments." The 2010 International Conference on High Performance Computing and Simulation, HPCS 2010.

Gantz, J. The Diverse and Exploding Digital Universe, IDC White Paper, 2008. http://www.emc.com/collateral/analyst-reports/diverse-exploding- digital-universe.pdf (accessed September 8, 2011).

Geambasu, R., Levy, A., Kohno, T., Krishnamurthy, A. and Levy, H.M. "Comet: An Active Distributed Key-Value Store." Proceedings of the Conference on Operating Systems Design and Implementation (OSDI), 2010.

Gilbert, S. and Lynch, N. "Brewer's Conjecture and the Feasibility of Consistent, Available, Partition-tolerant Web Services." *SIGACT News* 33 (2) (2002): 51–59.

Gladinet. Gladinet Cloud Storage Access Platform & Solutions, 2011. http://www.gladinet.com (accessed September 8, 2011).

Halevi, S., Harnik, D., Pinkas, B. and Shulman-Peleg, A. "Proofs of Ownerhip in Remote Storage Systems." 18th ACM Conference on Computer and Communications Security (CCS 2011), ACM SIGSAC, Chicago, Illinois, 2011,

Hanover, J. and Dunbrack, L.A. U.S. Health Industry Provider 2011 Top 10 Predictions: Solutions Providers' Need for ARRA and Healthcare Reform, IDC Health Insights, HI226422. January, 2011. http://www.idc-hi.com/getdoc.jsp?containerId=HI226422 (accessed September 8, 2011).

Harnik, D., Pinkas, B. and Shulman-Peleg, A. "Side Channels in Cloud Services: Deduplication in Cloud Storage." *IEEE Security & Privacy* 8 (6) (2011): 40–47.

INCTIS. Object-based Storage Device Commands (OSD), ANSI INCITS 400-2004, T10 Committee of the International Committee for Information Technology Standards, 2004. http://www.t10.org/cgi-bin/ac.pl?t=f&f=osd-r10.pdf (accessed September 8, 2011).

Isard, M., Budiu, M., Yu, Y., Birrell, A. and Fetterly, D. "Dryad: Distributed Data-Parallel Programs from Sequential Building Blocks." European Conference on Computer Systems (EuroSys), Lisbon, Portugal, March 21-23, 2007.

Jacobson, V., Smetters, D.K., Thornton, J.D., Plass, M.F., Briggs, N.H. and Braynard, R.L. "Networking Named Content." Proceedings of the 5th International Conference on Emerging Networking Experiments and Technologies (CoNEXT '09), ACM, New York, 2009.

Jclouds. jclouds Inc., A multi-cloud library, 2011. http://www.jclouds.org (accessed September 8, 2011).

Jungle Disk. 2011. https://www.jungledisk.com (accessed September 8, 2011).

Kolodner, E.K.., Tal, S., Kyriazis, D., Naor, D., Allalouf, M., Bayuh, E., Bonelli, L., Brand, P., Eckert, A., Elmroth, E., Gogouvitis, S., Harnik, D., Hernandez, F., Jaeger, M.C., Lopez, J.M., Lorenz, M., Messina, A., Shulman-Peleg, A., Talyansky, R., Voulodimos, A and Wolfsthal, Y. "A Cloud Environment for Data-intensive Storage Services", submitted to 3rd IEEE International Conference on Cloud Computing Technology and Science, CloudCom 2011.

Libcloud. Apache Libcloud: a unified interface to the cloud, The Apache Software Foundation, 2011. http://libcloud.apache.org (accessed September 8, 2011).

Marks, H. "IBM Wins Supercomputing Storage Challenge with GPFS-SNC." Network Computing For IT by IT, UBM TechWeb, November 23, 2010, http://www.networkcomputing.com/servers-storage/ibm-wins-supercomputing-storage-challenge-with-gpfs-snc.php (accessed September 8, 2011).

Microsoft. Microsoft Windows Azure Storage Service Level Agreement (SLA), Microsoft Corporation, 2010. http://www.microsoft.com/download/en/details.aspx?displaylang=en&id=6656 (accessed September 8, 2011).

Nirvanix. Nirvanix Service Level Agreement, Nirvanix Inc., 2011. http://www.nirvanix.com/sla.aspx (accessed September 8, 2011).

Open Group. Distributed Audit Service (XDAS), Preliminary Specification, The Open Group, January, 1998. https://www2.opengroup.org/ogsys/jsp/publications/PublicationDetails.jsp?catalogno=p441 (accessed September 8, 2011).

OpenStack. OpenStack Object Storage, OpenStack Cloud Software, 2011, http://openstack.org/projects/storage (accessed September 8, 2011).

Rabinovici-Cohen, S., Factor, M., Naor, D., Ramati, L., Reshef, P., Ronen, S., Satran, J. and Giaretta, D.L. "Preservation DataStores: New Storage Paradigm for Preservation Environments." *IBM Journal of Research and Development* 52 (4) (2008): 389–400.

Rochwerger, B., Ragusa, C., Breitgand, D., Clayman, S., Epstein, A., Hadas, D., Levy, E., Loy, I., Maraschini, A., Massonet, P., Munoz, H., Nagin, K., Toffetti, G., Tordsson, J. and Villari, M. "RESERVOIR—When One Cloud is Not Enough." *Computer* 44 (3) (2011) 44–51.

Rochwerger, B., Breitgand, D., Levy, E., Galis, A., Nagin, K., Llorente, I.M., Montero, R., Wolfsthal, Y., Elmroth, E., Caceres, J., Ben-Yehuda, M., Emmerich, W., and Galan, F. "The RESERVOIR Model and Architecture for Open Federated Cloud Computing." *IBM Journal of Research and Development* 53 (4) (2009): 535–545.

Schmuck, F. and Haskin, R. "GPFS: A Shared-disk File System for Large Computing Clusters." Proceedings of the 1st USENIX Conference on File and Storage Technologies (FAST '02), Berkeley CA, USENIX Association, 2002.

Silk, D. "CDMI and Cloud Federation", SNIA Storage Developer Conference. September 2010. http://www.snia.org/sites/default/files2/sdc_archives/2010_presentations/wednesday/DavidSlick_CDMI_Cloud_Federation.pdf.

SNIA. eXtensible Access Method (XAM), XAM v1.0.1:2009, SNIA XAM Initiative, http://www.snia.org/forums/xam/technology/specs (accessed September 8, 2011).

SNIA. Cloud Data Management Interface (CDMI), CDMI Version 1.0.1h:2011, SNIA Cloud Storage Technical Work Group, 2010. http://www.snia.org/cloud (accessed September 8, 2011).
SNIA. Managing Private and Hybrid Clouds for Data Storage, SNIA, January 2010. http://www.snia.org/sites/default/files/Private-Hybrid CloudWhitePaper_0.pdf (accessed September 8, 2011).
VISION Cloud. Introducing VISION Cloud, 2011. http://www.Vision cloud.eu/ dossier.php?s=30 (accessed September 8, 2011).

CHAPTER FOUR

MIGRATING LEGACY APPLICATIONS TO THE SERVICE CLOUD PARADIGM: THE REMICS PROJECT

PARASTOO MOHAGHEGHI, FRANCK BARBIER, ARNE J. BERRE, BRICE MORIN, ANDREY SADOVYKH, THOR SÆTHER, ALEXIS HENRY, ANTONIN ABHERVÉ, TOM RITTER, CHRISTIAN HEIN, MICHAŁ ŚMIAŁEK

The objective of the REMICS project is to develop model-driven methods and tools that support organisations in the migration of their legacy applications to the service cloud paradigm. The migration process consists of understanding the legacy system in terms of its requirements, architecture and functions, designing a new service-oriented application that provides the same or better functionality and quality of service, and verifying and implementing the new application in a cloud computing platform suitable for the purpose. To support the migration process, REMICS will enhance the Object Management Group's model-driven architecture methodology with specific methods, metamodels and tool support, including extensions to the Knowledge Discovery Metamodel, patterns and transformations for service-oriented architecture, cloud computing, requirements specification, model-driven interoperability, and extensions for testing and runtime management of services. REMICS tools and techniques are supported by the REMICS methodology for migration, which is based on agile method engineering principles. The REMICS consortium consists of eight research partners and two industrial users. The project started in September 2010 and runs for three years. REMICS is supported by the European Commission in the Objective 1.2 of FP7 Call 5.

1. Introduction

The Software as a Service (SaaS) paradigm has become more and more popular, enabling flexible payment models and moving the infrastructure management costs from consumers to service providers. However, building an SaaS system from scratch may require huge investments in time and effort. Therefore, it is sometimes necessary to build the system by reusing and modernising the legacy applications of an organisation. However, migrating legacy systems is neither automatic nor easy.

REMICS (REuse and Migration of legacy applications to Interoperable Cloud Services (http://remics.eu) is a research project supported by the European Commission that started in 2010 and will run for three years. The project's main objective is to develop model-driven methods and tools that support organisations with legacy systems to modernise them according to the service cloud paradigm. In our view, the cloud computing paradigm enhances the perception of IT companies as service providers. We therefore talk of the *service cloud paradigm* that stands for the combination of cloud computing and Service-Oriented Architecture (SOA) for the development of SaaS systems.

The REMICS consortium originally consisted of:

- SINTEF, research organisation and project coordinator from Norway
- SOFTEAM, SME (Small and Medium Entreprise) and provider of the Modelio case tool (http://www.modeliosoft.com/) from France
- Fraunhofer Institute for Open Communication Systems (FOKUS), research organisation from Germany
- TECNALIA, research organisation from Spain
- Netfective Technology, SME and provider of the BLU AGE® (http://www.bluage.com/) application generator from France
- DI Systemer AS, software developer and vendor within the Enterprise Resource Planning (ERP) / Accounting/ Customer Relationship Management (CRM) domain from Norway
- DOME Consulting & Solutions, software developer and provider of services to the travel industry from Spain.

After the first year, the project is extended to include additional partners from the "Enlarged EU". The new partners joining the project from September 2011 are:

- Warsaw University of Technology from Poland (WUT)
- University of Tartu from Estonia (UT)

- Institute of Information and Communication Technologies – Bulgarian Academy of Sciences from Bulgaria (IICT-BAS).

The REMICS project has already been presented in several publications such as Mohagheghi et al. (2010) and Mohagheghi and Sæther (2011). This chapter provides a more detailed presentation of the project and presents some of the latest results.

The remainder of this chapter is structured as follows. We discuss motivation and research challenges in Section 2. Section 3 provides an introduction of the technological approach of REMICS in five research areas and presents some preliminary results. Section 4 presents planned contributions to the standards while Section 5 summarises the chapter and includes conclusions.

2. Motivation and Challenges

The need for agility for IT infrastructures and information systems forces organisations to move to up-to-date technologies and applications. Year after year, the explosion of costs generated by the maintenance of old systems has also become a critical issue. The idea behind cloud computing is to provide on-demand network access to a shared pool of computing resources. Promises are many, among them cost-savings, agility, improved service availability and scalability. Benefits offered by new technologies and development paradigms such as SOA and cloud computing motivate companies to evaluate whether they can take advantage of these benefits while avoiding redeveloping their applications. However, modernisation often hides major challenges such as:

- **The oldness of the technologies to be reversed**: legacy systems often reveal very specific software architectures with homemade tuning. For example, in REMICS the industrial cases have parts with heterogeneous technologies like COBOL on one side, Delphi on the other and even .NET/C# and PL/SQL parts that require very specific platforms to work. (COBOL itself is subject to varied dialects. Supersets also exist based on a richer COBOL, often creating stronger architectural dependencies.) Moreover, old software often includes weakly separated data formats and poor code architectural features and infrastructures (e.g. CICS for COBOL).
- **The huge volume of code and data to be reversed**: volume is often a barrier to modernisation. Even though there are tools that

promise automatic parsing and reverse engineering, the process still needs effort from personnel that is costly and also has a psychological facet.

- **The absence of in-depth knowledge and the sharing of knowledge between previous and current employees**: After a few years and some updates, in-depth knowledge on the legacy system is often lost and the modernisation activity often meets some unexpected challenges.
- **The risks associated with new technologies while the business must go on as usual**: new technologies are often not well tested which is a major risk. It is also expected that the modernised system should provide the same functionality as the old one with even better performance. Convincing and training people to use new technologies are key aspects of the process.

There are several software engineering challenges also related to SOA and cloud computing (Mohagheghi and Sæther 2011):

- The SOA paradigm implies the breakdown of architecture into coarse-grain components providing business services. These components should be loosely coupled and have clear functionality.
- It should be possible to reconfigure services or compose them in new ways to support new tasks.
- Modernising the architecture of a legacy system to SOA requires developing models of new business processes, identifying services both top-down and bottom-up, and then deciding how to transform the current architecture to the migrated one, inevitably in a semi-automated way.
- Cloud computing technologies put architecture constraints on the client applications which will impact the migration process of an existing application to the cloud. Componentisation of architecture should enable scalability and thus make possible the multiplication of a number of instances of the same component. On the other hand, when creating a cloud application, it is necessary to ensure that the application can run on a set of low performance resources, to cope with network latencies in a loosely coupled environment. The decomposition and decoupling aspect is thus similar to SOA, with the additional scalability concerns.

Babar and Chauhan (2011) mention that there is also a need for new architecture evaluation methods since the existing methods do not emphasise cloud quality attributes such as scalability and accessibility.

The REMICS technological approach is developed to address the above challenges, and is based on Model-Driven Engineering (MDE) techniques where models are considered as primary artifacts in the software development life cycle. Models promote abstraction and hiding platform details and improve comprehension of the old system. SOA improves the architecture of the system by modularisation and increases interoperability with other systems. The MDE approach allows automation by transformations while humans are also inevitably involved in many tasks. Finally our focus on cloud computing allows for the investigation of the impact of this new technology on the performance and scalability of applications.

The project includes two industrial pilot cases in the domains of ERP/Accounting/CRM and travel industry. Some of the challenges of the cases are described in Mohagheghi and Sæther (2011). The new partners are also in contact with companies with cases of legacy systems to modernise. This includes an online Corporate Banking application to be deployed in cloud (by WUT), a transport system for migration to the SaaS paradigm (by IICT-BAS), and Online Transaction Processing (OLTP)/ Online Analytical Processing (OLAP) applications to be migrated to cloud environments (by UT). OLTP systems in the form of three-tier systems and OLAP systems are legacy scientific applications on the desktop. Thus, the REMICS approach will be evaluated in several domains.

3. The REMICS Technological Approach

Fig. 4.1 below depicts the technological approach of REMICS. Modernisation in this process combines reverse engineering of legacy applications with forward engineering to the desired platforms. In the remainder of this section, we discuss the steps in detail.

The modernisation of legacy applications starts with understanding the functionalities and organisation (particularly the architecture) of a legacy application in the “recover” activity. The reverse engineering process extracts elements from legacy code and data description, rendering them into a Platform Specific Model (PSM). The Object Management Group’s (OMG) Knowledge Discovery Metamodel (KDM, http://omg.org/ technology/documents/modernization_spec_catalog.htm) is the candidate support for representing PSMs using the Abstract Syntax Tree Metamodel (ASTM) as the sub-support for the precise representation of programming language features, including odd characteristics. ASTM was released in January 2011 to fill a gap in the daily use of KDM for legacy code manipulation especially.

Fig. 4.1. REMICS technological approach

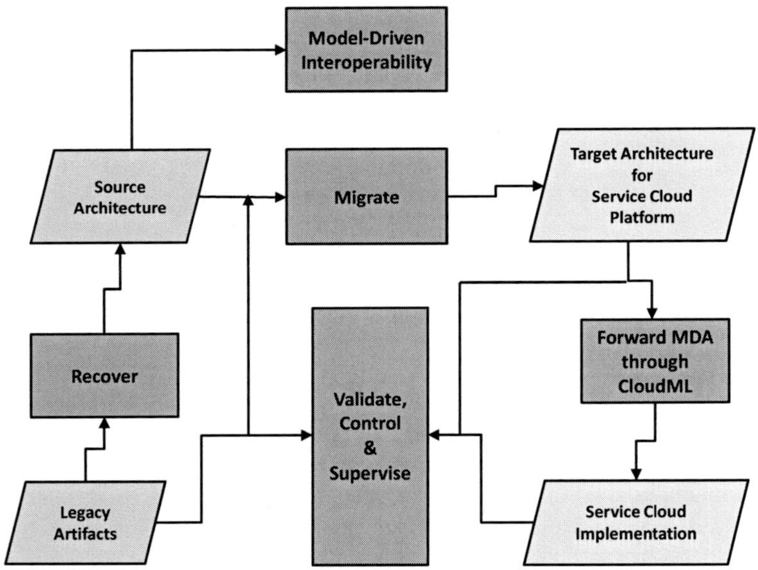

3.1. Recover

As an illustration, fig. 4.2 shows how a simple COBOL snippet may be translated into an ASTM model. This model is semantically poor in the sense that no information about persistence issues, inputs/outputs issues, platform dependence issue etc. is presented in the model. The REMICS recovery process aims at enriching these raw models and transforming them through many codified steps to rich models that are bases for transformation to Unified Modeling Language (UML) models.

Fig. 4.2. ASTM model corresponding to a "DATA DIVISION" section

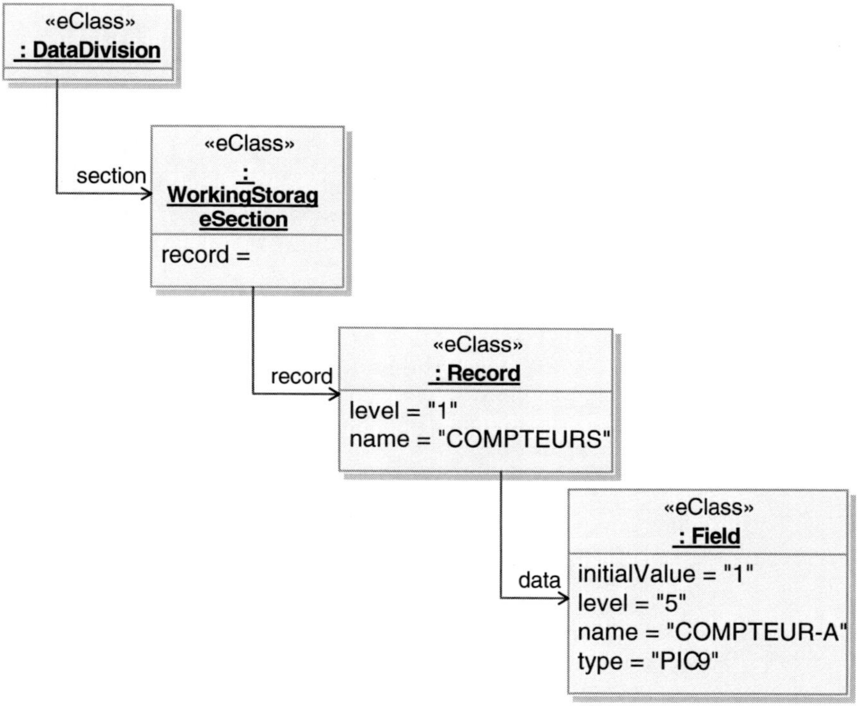

ASTM is composed of GASTM (Generic Abstract Syntax Tree Metamodel) and SASTM (Specific Abstract Syntax Tree Metamodel). While GASTM allows the representation of the code without any language specificity, SASTM allows the representing of code peculiarities by developing metaclasses/metarelationships on the top of GASTM. Due to the absence of a clear and formal interrelation specification between KDM and ASTM, REMICS has developed a metamodel named EKDM (Extension of KDM). This means keeping both the support for data modelling, User Interface (UI) modelling, platform modelling (for example legacy operative system calls or legacy platform service usage) provided by KDM and the level of precision for old-fashioned language modelling allowed by ASTM.

Having EKDM as a set of additional metaclasses and metarelationships is however not enough—one also needs a set of predefined model transformations along with a well-codified transformation process to move from one model to another. This process and the set of transformations is the heart of the BLU AGE REVERSE engine, which is used in REMICS in the "recover" activity.

Knowledge discovery has so far often been limited to reverse engineering of legacy code, with the business process and rules recovery being poorly addressed. One objective of EKDM is to address this gap and provide practical methods and tool support for recovering business value information from the legacy system artifacts. For that, KDM includes a conceptual package aligned with another (standardised) metamodel named Semantics of Business Vocabulary and Rules (SBVR, http:// www.omg.org/spec/SBVR/1.0/).

Fig. 4.3 below summarises the "recover" activity and shows the outputs. The final steps of the "recover" activity are composed of transformation from KDM/ASTM models into UML Platform Independent Models (PIM). During this activity, no reorganisation or changes are done, while the ultimate goal is to discover and understand the legacy application in multiple models representing the functionalities, architecture, behaviour and business rules.

Fig. 4.3. Steps in the "recover" activity

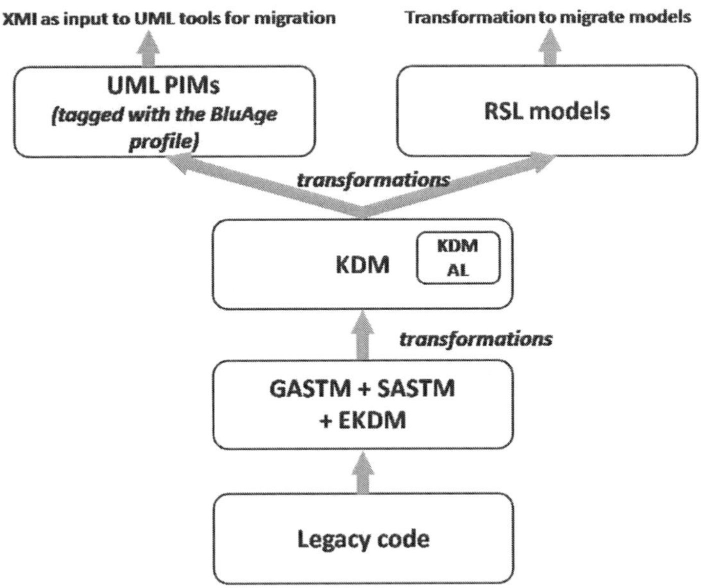

The extended REMICS project will also bring the requirements factor to the models recovered from the legacy systems. For this purpose, the

Requirements Specification Language, RSL (Śmiałek et al. 2010), will be applied as an intermediate language between the "recover" and "migrate" steps. This language will allow for recovering requirements-level models supplementing the architectural-level models, translated from the information captured through KDM. The KDM metamodel will therefore be extended with elements for representing the recovered Application Logic (AL). These KDM additions (together with the overall KDM metamodel) will be used for the automated transformation into RSL.

Requirements in RSL (at the Computational Independent Model or CIM level) are represented through user scenarios with precise syntax and semantics. For the RSL-related extensions, a tooling framework will be integrated into the REMICS tool suite. This framework will build upon the results of the ReDSeeDS project (6FP ICT Objective 1.2, http://www.redseeds.eu/) where the RSL editor and transformation engine were developed by WUT (Śmiałek et al. 2010). The idea is to use RSL to generate architectural-level models in SOA in the "migrate" activity when possible and test cases since the application logic (observable behaviour) level information is captured.

3.2 Migrate

Models from the "recover" activity are the starting point for the "migrate" activity depicted in fig. 4.4 below. During this activity, the new architecture of the migrated system is built by applying specific SOA and cloud computing patterns and methods like architecture decomposition, legacy components' wrapping, and legacy components' replacement with newly discovered cloud services. However, the recovered models have to be drastically re-factored. The monolithic components have to be decomposed and decoupled in order to obtain an autonomous SOA architecture. Newly discovered online services should be integrated into design as well.

The re-architecting (and thus re-expressing) happens by means of the *Service-oriented architecture Modeling Language* SoaML (Sadovykh et al. 2010; Elvesæter et al. 2011; http://www.omg.org/spec/SoaML/1.0/Beta2/). SoaML is a UML profile developed based on contributions from an international team, including the FP7 SHAPE project led by SINTEF (http://www.shape-project.eu/), and is being standardised by OMG. SoaML can be used for architecture level modelling, or as part of a Model-Driven Architecture (MDA) process, starting with a business model and transitioning through logical and physical models, resulting in technology implementation. This will be a foundation for the service models in

REMICS. In addition, the recovered RSL models will be transformed into the target architectural models at the PIM level.

Fig. 4.4. An overview of the “migrate” activity

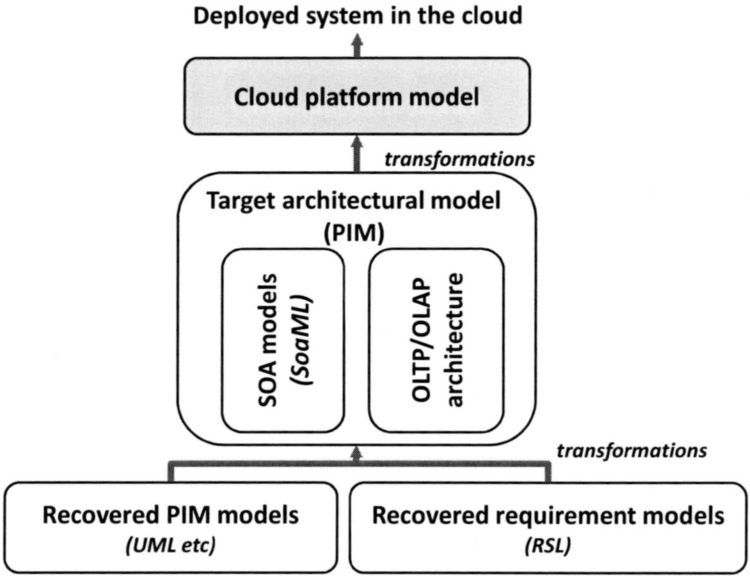

REMICS will extend SoaML to address the specific architectural patterns and model driven methods for architecture migration. The Modelio UML modelling and profiling toolset provided by SOFTEAM is enhanced to support the needs of REMICS, as well as in component identification and architecture refactoring. The tool is intended to manage the integration mechanisms between the different tools involved in the process of modernisation.

In order to provide model-driven support for deployment, REMICS develops *PIM4Cloud* as a platform independent and technology neutral standard for cloud computing. A *Cloud Platform Generator* will manage specific code generators and descriptor generators for cloud platform services expressed within PIM4Cloud models.

As part of the “migrate” activity, we have identified our requirements towards cloud computing platforms and have evaluated some of the well-known platforms. We focused on the identification of services that cloud computing platforms offer but also on the constraints brought by platforms. Platform as a Service (PaaS) level technologies such as

Windows Azure (http://www.microsoft.com/windowsazure/) tend to impose a specific architecture of deployed applications. Our experiments on the Infrastructure as a Service (IaaS) level platforms have shown their impact on application architecture remains limited while they can handle deployment issues in the cloud. However, these platforms deliver a service level below that of PaaS platforms. Currently, Amazon Elastic Compute Cloud (Amazon EC2, http://aws.amazon.com/fr/ec2/) acts as a de-facto standard in the IaaS market and other providers try to be compatible with it, reducing the risk of having too diverse solutions. Amazon EC2 seems to be a proper solution for moving our case applications to the cloud and is also suitable for hybrid clouds.

3.3 Validate, Control and Supervise

The "validate, control and supervise" activity groups the technologies that are dedicated to ensure the correctness of the migrated legacy system.

First, we take advantage of static model analysis techniques by defining and collecting measures at the model level. In Mohagheghi and Dehlen (2009), state of the art regarding model metrics presented. Work is on-going in identification of appropriate measures for SOA and cloud computing. Some measures from the object-oriented modelling paradigm can be adopted here:

- Number of Services - is a simple count of all services in a system. It is a size metric that only considers the pure count of services in a system;
- Coupling of Service - two components are coupled if and only if at least one of them acts upon the other;
- Inter-Service Coupling - is the count of channels between the two services;
- Well-chosen granularity - service granularity is measured by the number of operations defined in a service.

The Metrino tool (www.modelbus.org/modelbus/index.php/metrino) developed by FOKUS is to be used for collecting the identified measures. Metrino is an integrated set of tools for the support of the analysis, verification and quality assurance of models based on the Object Constraint Language (OCL, http://www.omg.org/spec/OCL/2.0/) and Structured Metrics Metamodel (SMM). The tool provides the functionality to manage and compute generated or user-defined measures and to visualise their computational results in appropriate charts.

Model-Based Testing (MBT) is an approach to validate specific properties of a system/model. This approach makes use of different methods and techniques such as generating test models and test data. The test model may be derived from the requirements (similar to the system model but with a different purpose) or it can be partially derived from the system model. Both methods will be investigated in REMICS considering the recovery of both system models and requirements of the legacy applications.

With respect to MBT, the Fokus!MBT tool is used in the assessment of the renewed applications. The MBT environment is depicted in fig. 4.5.

Fig. 4.5. Model-based testing in REMICS

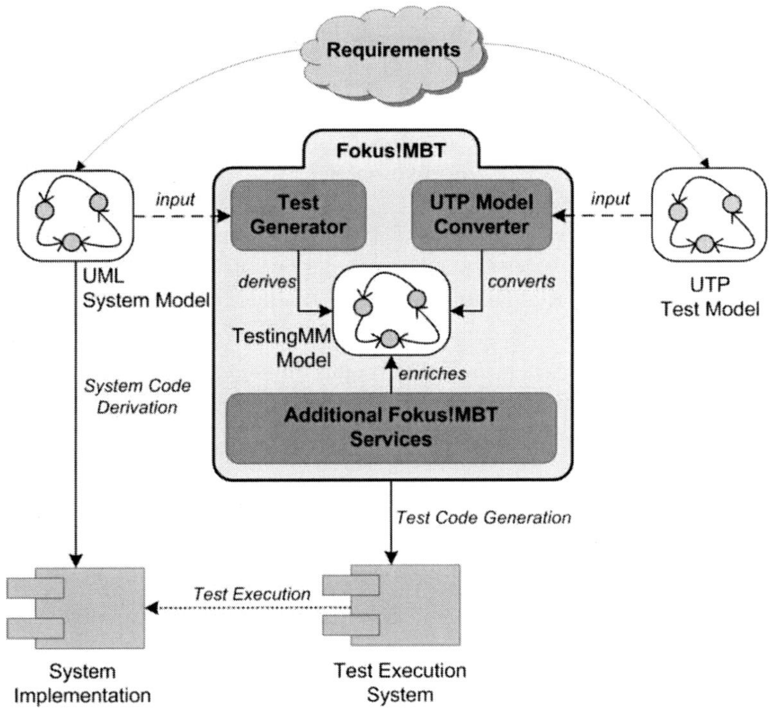

REMICS investigates existing test notations such as the UML2 testing profile (UTP, http://www.omg.org/technology/documents/profile_catalog.

htm) for their application to SOA and cloud computing and if needed will extend them. The profile is intended to support an effective, efficient and automated testing of system implementations according to their UML models. In the figure, TestingMM stands for the testing metamodel as the basis for test models.

Another research area in this activity is managing models at runtime so they satisfy certain characteristics. Due to the very recent emergence (and thus weak maturity) of the notion of Models@Runtime (Ballagny et al. 2009; MRT 2011), REMICS aims at exploring this notion in order to provide a proof of concept. This proof of concept aims to measure the added values of runtime verification compared to MBT or static model analysis that occur at design time. Specifically, cloud platforms require, compared to traditional SOA, more manageability support, even self-manageability. The ability to transparently scale up, to have fail-in-place components without disturbances illustrates manageability. With clouds, developers must have at their dedicated APIs (Application Programming Interfaces). The REMICS view offers actions on Models@Runtime as management functions. Developers may develop self-management functions in their cloud services (self-healing for fault resilience, self-optimisation for scalability, etc.).

3.4 Interoperate

When a legacy system has been migrated to the cloud, it directly competes with a plethora of other applications which might provide similar services, with different costs or Quality of Service (QoS). It is thus important to support the modernised application with advanced interoperability mechanisms so that it can evolve in a win-win relationship in the cloud ecosystem. In addition, there may be a need to interoperate with third party systems in providing services or compose services to new ones offered to the users.

Since the recovery and migration steps intensively rely on models and MDE techniques, REMICS applies MDE techniques to provide support for interoperability as well. There are two scenarios for interoperability in REMICS industrial cases:

- The first scenario is the need for **data interoperability**. The migrated systems need to interoperate with different clients and providers that have services based on other data schemata than the legacy application. For example in the DI Systemer case, exchange of data between services may be based on the XBRL Global Ledger (GL) standard (http://www.xbrl.org/GLTaxonomy/). XBRL GL indeed offers an extensible, flexible, multi-national solution that can exchange the data required by finance workers, accountants, and creditors.

- The second scenario is the need for **service interoperability**. If designers want to replace a service with another similar service available in the cloud, it is very likely that the external service does not rely on the same protocol (the sequence of messages interacting with the service) than the former service. For example, some messages need to be split in two, while some messages need to be merged. The order of the sequence might also be slightly different.

Since SoaML is central in REMICS, we use UML class diagrams for data interoperability and UML behavioural diagrams for service mediation, which are *de facto* compliant with SoaML and provide the necessary expressiveness.

Regarding data interoperability, REMICS is developing a tool for semi-automatic inference of (semantic) mappings between two data models. Fig. 4.6 below gives an overview of the proposed data exchange framework.

The shadowed box in the centre of fig. 4.6 (to which we refer to as *MDI Transform Engine*), is the core part of the framework, where all the specifications and transformations take place at the platform-independent level. The elements outside the shadowed box (source and target schemas, their transformation, the source instance, and the generated target instance) represent platform-specific models and transformations.

At the heart of the transformation is a mapping metamodel (depicted by Mapping MM in the figure) for specifying the schema mappings (Mappings Specifications in the figure) in an intuitive way at the platform-independent level. Based on the informal schema mappings provided by the domain experts, the designer formalises these mappings by instantiating a model conforming to the mapping metamodel. Furthermore, our framework provides support for the automatic generation of executable mappings used at run-time to execute the transformation of the Source Model (source data instances) to the Target Model (target data instances). This way, the user needs to focus on the transformations at the platform-independent level, with the actual data exchange being fully automated.

Fig. 4.6. Model-driven data exchange approach

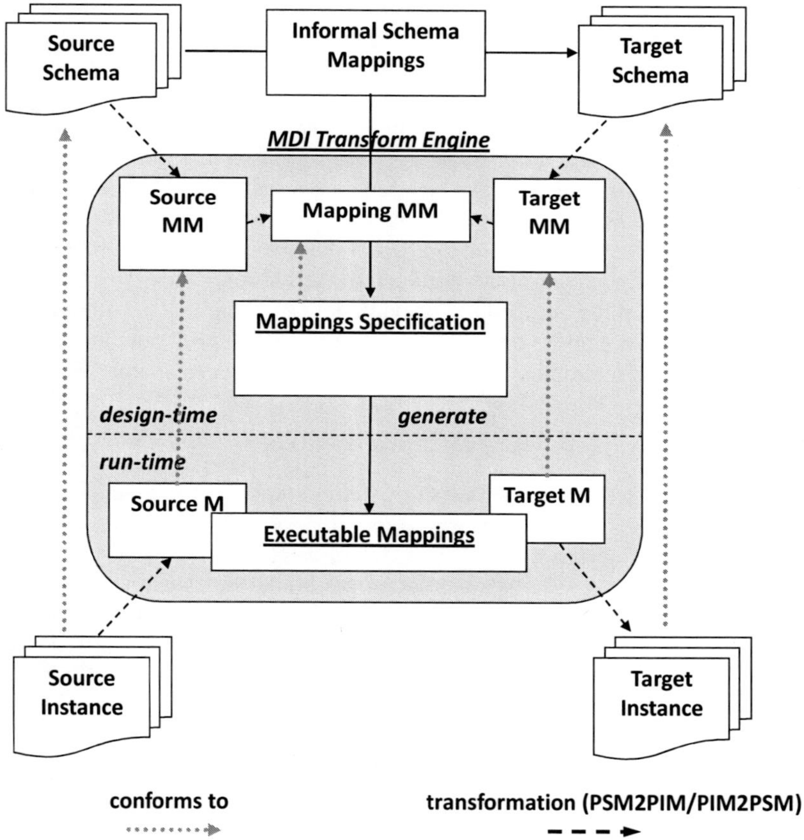

For managing the second scenario, the developer has two choices: either using this new service as-is and directly modifying all the client services of the modernised system, or providing interoperability mechanisms, which will bridge the gap between the former service and the new one. The first solution has the advantage of having no overhead at runtime: clients are modified so that they can directly and efficiently communicate with the new service. The drawback is the need for updating applications that are using the service, which may be cumbersome if there are several replacements. The second solution is based on "service mediation," which will be investigated in REMICS. The client services remain unchanged by inserting some "glue" between clients and the new service. This implies a slight overhead (indirection, extra processing, etc.), which is however not problematic in the kind of applications addressed in REMICS. The mediation services are expressed through finite state machine models and the needed concepts are under development.

Innovations in REMICS regarding service mediation are twofold. First, REMICS adopts a MDE approach to generate mock-ups, skeletons and simulators that allow designers to rapidly prototype and iteratively define mediators, with no need to use the concrete services (which can be costly to use if they are pay-per-use services). Second, REMICS relies on Models@Runtime as executable models to implement and deploy mediation services. A software library deployed as a self-contained component in the modernised application acts as a virtual machine in charge of executing embedded models. The dynamicity offered by embedded models makes it possible to dynamically deploy new mediators so that a client can seamlessly switch from one provider to another one.

3.5 REMICS Methodology

The goal of the REMICS methodology is to identify the challenges in the different activities of the modernisation process and to provide a tool-supported guideline based on REMICS tools and technologies to support users in this process. This is work in progress and the first release (to be published in September 2011) will contain the overall description of the methodology, including methodology principles, description of main activities (Requirements, Recover, Migrate, Testing, Interoperability, etc.), and some details of the initial activities.

As part of this work, we have done a comprehensive state of the art analysis related to all REMICS research areas and the results are already published on the REMICS website (D2.1, http://remics.eu/public deliverables). Regarding methodology, there are several service engineering methodologies such as:

- Service Architecture Engineering (SAE), which extends the reference model by OASIS for SOA (Butler 2007)
- Service-Oriented Analysis and Design (SOAD), developed by IBM (Zimmermann et al. 2004)
- Service Oriented Development of Applications (SODA), developed by Gartner Research and with emphasis on reusability (Gartner n.d.)
- Service-Oriented Modeling and Architecture (SOMA) (Arsanjani et al. 2008), also developed by IBM

- The methodology developed by the MOMOCS project (MOdel driven Modernization of Complex Systems; http://www.momocs.org/), a completed collaborative project developed with the support of the European Commission in the context of the FP6. It is called XIRUP and has four phases: preliminary evaluation, understanding, building and migration. The methodology however has no focus on SOA
- Service Migration and Reuse Technique (SMART); developed by Carnegie Mellon Software Engineering Institute (SEI). It is defined as a process to help organisations make initial decisions about the feasibility of reusing legacy components as services within a SOA environment.

SMART and XIRUP are used in our methodology for identifying necessary steps in the migration process. Other methodologies focus more on developing services from scratch.

We have also analysed existing cloud computing platforms and cloud deployment models as discussed in Section 3.2. However, there is no methodology that guides developers through the process of selecting the platform that best meets their needs.

Many companies are using agile methods for software development. Some agile techniques may especially be useful in order to reduce risks and provide better software quality in migration projects; for example code refactoring and continuous integration and testing. We therefore study which agile techniques may be integrated in the REMICS methodology.

REMICS methodology will initially be described by using the Eclipse Process Framework (EPF, http://www.eclipse.org/epf/). EPF is an open source software process engineering tool developed over Eclipse. The tool can be downloaded for free and will run on any capable platform. While we take advantage of EPF at this stage, work on defining a more dynamic and easy to use method engineering framework is going on. It is assumed that such a framework can in particular provide better support for SMEs. REMICS is involved in defining and standardising such a framework, which has started within the SEMAT community (http://www.semat.org/bin/view). A Request for Proposal (RFP) with the title “A Foundation for the Agile Creation and Enactment of Software Engineering Methods" led by SINTEF was approved on June 24, 2011. SEMAT looks at methods as a composition of a number of “practices”. Practices are mini-methods with a clear beginning and end; the intention is that the practice will deliver something of value. Practitioners are free to mix and match the practices they feel are appropriate for their work. We will contribute to the development of this framework and tools supporting it.

4. REMICS Contributions to Standards

REMICS will focus on open source metamodels and models with an emphasis on open models for standards based on open interfaces. We will typically provide EMF (http://www.eclipse.org/modeling/emf/) based versions of the models representing the standards being worked on and provided in REMICS. This will allow for easy access to the standards and easy implementation of these by multiple vendors. In order to instrument the migration process the REMICS project will integrate a large set of metamodels and will propose several dedicated extensions as depicted in Fig. 4.7 below (In Gray Color).

Fig. 4.7. REMICS contributions to standards

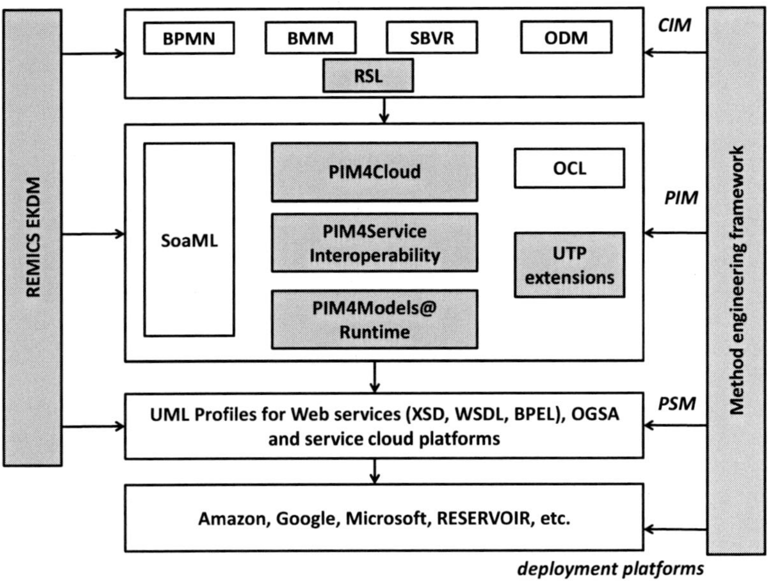

At the CIM level, RSL will be proposed as a standard for specifying requirements. Based on the project experience in the recovery activity, the consortium will propose extensions to KDM (Extended KDM—EKDM) required for fulfilling the REMICS methodology. PIM4Cloud for modelling implementation in cloud will be proposed for standardisation through an OMG RFP process or similar. Other extensions to standards for the modeling of SaaS paradigm on the platform independent level to be considered will be PIM4ServiceInteroprability, PIM4Models@Runtime and UTP extensions. Finally, the methodology research includes participation in defining a language for the software engineering domain for agile method engineering.

5. Summary and Conclusions

This chapter has described the objectives of the REMICS project and presented preliminary research results. The REMICS approach described in Section 3 includes five research areas that together address the challenges of migration as introduced in Section 2 of this chapter:

- **Recover** supports the efficient extraction of information from the legacy system artifacts and presents these in models.
- **Migrate** supports modernising the software architecture into SOA and extending the SOA models with cloud specific elements.
- **Interoperate** supports interoperability in a systematic way between services and data after modernisation.
- **Validate, Control and Supervise** supports robust and automated testing strategies that guarantee that the new system provides the same functionality along with the same or even better QoS. It will also help to manage the SaaS application at runtime using the available models.
- **REMICS methodology** integrates the methods and tools developed within REMICS in an Integrated Development Environment (IDE). It also provides a step-by-step guide on how to use them.

Our approach to migration preserves and capitalises on the business value engraved in legacy systems to gain flexibility brought by service clouds. The goal is to lower the cost of service provision and shorten the time-to-market for providing services. The REMICS approach is developed in close collaboration with industrial users and their cases are used in identifying challenges, evaluating the feasibility of the approach and providing feedback to tool and technology providers. We have therefore identified their requirements and used them as basis for evaluation. We have also identified other success criteria for the project related to the European Commission objectives.

For the best exploitation of the results inside and outside of the consortium, the project focuses heavily on standardisation. We presented our planned contributions to standards in Section 4 of this chapter.

There are several challenges in this work:

- **Extending SOA models with peculiarities of the cloud computing paradigm**: the notion of cloud computing groups together several heterogeneous forms, both in terms of the services provided (Service, Platform and Infrastructure) and the types of implementation (Private, Public and Hybrid). Standardisation work on cloud platforms is at the moment too diverse and performed by several entities from different viewpoints.
- **Automating the migration process with model transformations**: The MDE approach is considered useful if we can automate the process to a high degree. The challenge is related to all aspects of our research; from extracting business value information from legacy systems to migration of the software architecture to SOA models, managing the interoperability of services, managing services at runtime and testing the migrated systems.
- **Developing a reusable and easy to use approach to migration**: legacy systems have many peculiarities and developing a general approach that fits all systems is not possible. However, our goal is to develop platform independent approaches and take advantage of well-established OMG MDA technologies to support reuse as much as possible.
- **Systematic evaluation of the results**: technology development requires systematic methods for evaluation that generate reliable results for the outside world. We therefore work on an evaluation plan based on various requirements to our solutions that is also part of this research.

The first versions of REMICS tools and methodology are released in September 2011 and we continuously update the project website with public deliverables, open source tools and demonstrations, and links to published articles.

Acknowledgements

This work has been funded by the European Commission through the REMICS project (http://remics.eu), contract number 257793, within the 7th Framework Program.

References

Arsanjani, A., Ghosh, S., Allam, A., Abdollah, T., Ganapathy S. and Holley K. "SOMA: a Method for Developing Service-Oriented Solutions". *IBY Systems Journal* 47 (3) (2008).

Babar, M.A. and Chauhan, M.A. "A Tale of Migration to Cloud Computing for Sharing Experiences and Observations". Proceedings of Software Engineering for Cloud Computing Workshop (SECLOUD '11) at ICSE 2011.

Ballagny, C., Hameurlain, N. and Barbier, F. "MOCAS: A State-Based Component Model for Self-Adaptation". SASO 2009, 206–215, 2009.

Butler, J. "The Architecture Component of the SAE Reference Framework for SOA". *CBDi Journal,* 2007. http://www.cbdiforum.com/secure/interact/2007-03/the_architecture_component.php (accessed January 31, 2011).

Elvesæter, B., Carrez, C., Mohagheghi, P., Berre, A.J., Johnsen, S.G. and Solberg, A. "Model-Driven Service Engineering with SoaML". Chapter in the book: *Service Engineering—European Research Results*, edited by S. Dustdar and F. Li, SpringerLink, 2011. http://dx.doi.org/10.1007/978-3-7091-0415-6 (accessed September 15, 2011).

Gartner, http://www.gartner.com/research/spotlight/asset_112573_895.jsp, (accessed June 15, 2011).

Mohagheghi, P. and Dehlen, V. "Existing Model Metrics and Relations to Model Quality". Proceedings ICSE Workshop on Software Quality (WOSQ '09), IEEE Computer Society, 39–45, 2009. http://dx.doi.org/10.1109/WOSQ.2009.5071555

Mohagheghi, P., Berre, A.J., Sadovykh, A., Barbier, F. and Benguria, G. "Reuse and Migration of Legacy Systems to Interoperable Cloud Services: the REMICS Project". Proceedings of MDA4Service Cloud2010, 2010. http://events.sti2.at/mda4ServiceCloud2010/; (accessed September 15, 2011).

Mohagheghi, P. and Sæther, T. "Software Engineering Challenges for Migration to the Service Cloud Paradigm". Proceedings of Services 2011: IEEE World Congress on Services, IEEE Computer Society, 507–514, 2011.

MRT 2011. http://www.comp.lancs.ac.uk/~bencomo/MRT11/; (accessed September 15, 2011).

Roman, D., Morin, B., Wang, S. and Berre, A.J. "A Model-driven Approach to Interoperability in B2B Data Exchange". In Advanced Result on MDI/SOA innovation workshop (at IWEI 2001), Stockholm, Sweden, March 2011. http://remics.eu/publications (accessed September 15, 2011).

Sadovykh, A., Desfray, P., Elvesæter, B., Berre, A.J. and Landre, E. "Enterprise Architecture Modeling with SoaML using BMM and BPMN—MDA Approach in Practice". 7th Central and Eastern European Software Engineering Conference, CECR 2010, Moscow, Russia, 2010.

SMART. http://www.sei.cmu.edu/library/abstracts/reports/08tn008.cfm; (acceeessed January 31, 2011).

Śmiałek, M., Kalnins, A., Kalnina, E., Ambroziewicz, A., Straszak, T. and Wolter, K. "Comprehensive System for Systematic Case-Driven Software Reuse". Lecture Notes in Computer Science (SOFSEM'10), 697–708, vol. 5901, 2010.

Abbreviations

AL	Application Logic
ASTM	Abstract Syntax Tree Metamodel
CIM	Computational Independent Model
EKDM	Extension of KDM
EPF	Eclipse Modeling Framework
IaaS	Infrastructure as a Service
KDM	Knowledge Discovery Metamodel
MDA	Model-Driven Architecture
MDE	Model-Driven Engineering
MDI	Model-Driven Interoperability
OCL	Object Constraint Language
OLAP	On-Line Analytical Processing
OLTP	On-Line Transactional Processing
OMG	Object Management Group
PaaS	Platform as a Service
PIM	Platform Independent Model
PSM	Platform Specific Model
QoS	Quality of Service
RSL	Requirements Specification Language
SaaS	Software as a Service

SME	Small and Medium Enterprises
SMM	Structured Metrics Metamodel
SoaML	Service-oriented architecture Modeling Language
UML	Unified Modeling Language
UTP	UML2 Testing Profile.

PART II:

CLOUD MANAGEMENT

CHAPTER FIVE

TOWARDS HOLISTIC CLOUD MANAGEMENT

JOHAN TORDSSON, KARIM DJEMAME, DANIEL ESPLING, GREGORY KATSAROS, WOLFGANG ZIEGLER, OLIVER WÄLDRICH, KLEOPATRA KONSTANTELI, ALI SAJJAD, MUTTUKRISHNAN RAJARAJAN, GEORGINA GALLIZO AND SRIJITH NAIR

Despite significant attention and substantial efforts both in industry and academia, cloud computing has yet to reach its full potential. Commonly stated obstacles for cloud adoption include confusion due to multiple delivery models (SaaS, PaaS, IaaS) and deployment scenarios (public clouds, private clouds, cloud bursting etc.). Other frequent concerns relate to risks with outsourcing, data legislation issues, inability to assess trust in external providers, etc. In addition to these obstacles to cloud computing as a concept, there are also technical thresholds in present cloud offerings, making cloud infrastructure provisioning and service lifecycle management tedious processes. We aim to address these issues by research in five main directions: cloud service lifecycle optimisation, adaptive self-preservation, self-management based on non-functional criteria, support for multiple cloud architectures, as well as market and legislative studies. The main outcome of our research is the OPTIMIS Toolkit, a set of flexible tools for service and infrastructure self-management.

1. Introduction

With its promise of close to infinite scalability, rapid provisioning, and a pay-per use charging model, cloud computing appears to be a compelling paradigm for service provisioning. However, as contemporary cloud offerings typically target functionalities too close to infrastructure, organisations must undertake significant efforts to manage their services in a cloud. In addition, the at times confusing cloud landscape with multiple models (SaaS, PaaS, IaaS) and architectures (private, public, federated, and bursted clouds) introduces additional thresholds to adoption, due to a lack of tools that abstract and hide the complexity. With its outsourcing nature, the cloud paradigm in itself introduces challenges related to trust and risk, as well as security and privacy of services and data, making organisations hesitate to migrate sensitive or mission-critical services to the cloud. For the rest of this chapter, we use the term "service" to describe an application that is accessed online by end users, and which is delivered by a cloud infrastructure provider through the use of one or more Virtual Machines (VMs).

We summarise our analysis of cloud computing obstacles in five high-level research challenges, discussed in more depth in previous work (Ferrer et al. 2011):

- **Service life cycle optimisation** concerns practices and tools to facilitate and optimise the construction (implementation and/or orchestration), deployment, and operation of cloud services.
- **Adaptive self-preservation** constitutes a wide range of self-management activities with increased autonomy as the ultimate goal, enabling fewer administrators to handle increasingly more services and larger infrastructures.
- Cloud and service **self-management based on non-functional management criteria** extends beyond traditional performance tuning to also incorporate non-functional aspects such as trust, risk, eco-efficiency, and cost (henceforth TREC) in self-management decision making.
- With **support for multiple cloud architectures**, a single abstraction can be used to construct a wide range of clouds. These include private clouds where one organisation provisions both services and the infrastructure needed by these; bursted clouds where organisations temporarily outsource non-critical workloads; federated clouds of collaborating cloud infrastructure providers that mutually share workloads; multi-cloud deployment, where a service owner negotiates with, and deploys a service across, multiple cloud infrastructures, with the negotiation process potentially being mediating by a cloud broker.
- Studies of **market and legislative issues** includes identification of new market roles and business models for clouds, as well as investigation of legal aspects related to the acquisition, transfer, and storage of service data.

Our main research objective is to make significant progress in these five areas and to incorporate our results into the OPTIMIS Toolkit. The OPTIMIS Toolkit is a set of software components aimed to simplify and optimise construction, deployment, and operation of services as well as operation of the virtualised infrastructure required to deliver these services.

The toolkit is not a turn-key solution but rather a set of software components that can be selected and composed for use in a variety of cloud deployment scenarios, as well as be customised through policies to meet different Business Level Objectives (BLOs).

The rest of this chapter is organised as follows: Section 2 outlines the working process and requirement engineering method used in the development of the OPTIMIS Toolkit. Section 3 describes the architecture of the toolkit, highlighting self-assessment aspects, service lifecycle management, and secure interconnection of multiple clouds. Section 4 presents our conclusions.

2. Process and Requirements

A multi-faceted project such as OPTIMIS with a wide range of overall objectives requires a structured working process. In the following, the used goal-oriented requirements process is outlined along with the overall requirements identified. The quality of requirements has a direct impact on the chances of success of the project. Therefore, an important task is to capture and structure the project requirements in order to ensure a number of qualities such as precision, completeness, consistency, traceability, and testability (OPTIMIS 2010a). In order to achieve these qualities, the requirements were elaborated based upon:

- A full project glossary capturing the domain vocabulary.
- A scenario approach to discover functional requirements of various stakeholders, using precise UML 2.0 sequence diagrams, both for normal and exceptional cases.
- A goal oriented methodology, to structure those scenarios into goal trees with a clear distinction between requirements under the responsibility of some component to design, assumptions on the existing environment, and the ability to check for consistency and completeness.
- The ISO-9126 standard to document all non-functional requirements.
- The definition of traceability mechanisms to reference requirements and also show the coverage between goals and scenarios, directly useful for verification and validation test cases.
- A methodology that ensures the different sources of the requirements (general, use cases, legal, business) are consolidated in a way that ensures requirements are specific, measurable, attainable, relevant, and time bound (SMART) to the software development.

These approaches were combined to build a requirements document that follows the IEEE-830 recommended practices for software requirements specifications. This document also helps understanding the project objectives and their rationale; the discovery of the main actors, their responsibilities and the information they need; high quality testing based upon the various comprehensive scenarios for quality assurance. The following OPTIMIS actors are identified:

- Service Provider (SP): Provides a service/application to the Customer.
- End User / Consumer: Entity (person or process) who is actually using the service provided by the SP.
- Infrastructure Provider (IP): The provider of the infrastructure resources, in IaaS fashion. It provides easy access to computing, storage and networking resources, according to the request from the SP to deploy and execute the application.
- Customer (C): Organisation who makes the contract with the SP, in order to request a service (or application).
- Broker (B): Sits between the infrastructure provider and the service provider. In addition to providing the abstraction layer of the individual IPs and their heterogeneous interfaces to the SPs, it provides service arbitrage, a matching engine, aggregation, and intermediation. The Broker can also offer other value-added services between the IP and the SP, such as benchmarking tools, performance prediction (based on monitoring historical data), security, SLA negotiation, among others. In a general case, multiple IPs play into the game (1-N relationship), such as in multi-cloud scenarios, but the Broker can also offer its services in the case of a 1-1 relationship between SP and IP.

2.1 Requirements from High-Level Objectives

A goal-oriented methodology has been used to elicit the functional requirements since it allows a clear separation of concerns between the requirements by relating them to distinct objectives. The methodology is based on the progressive refinements of abstract and general objectives into more concrete goals and requirements that lead to the identification of the main entities with their relationships and of the main agents responsible for the satisfaction of the requirements. The methodology is based on the following models:

- **Goal** model, capturing higher level objectives and more operational ones, connected by refinement links. Figure 5.1 provides a summary of the global goal model, its underlying properties, and the operational measures to satisfy these properties.

- **Object** model, giving a structured view of the pertinent vocabulary necessary to express the goals. This model is expressed using the classical UML class diagram with entities, relationships, attributes, cardinalities, etc.
- **Agent** model, gives a view of all the requirements under the responsibility of each agent. This means that each agent instance in the system (end-user, SP, IP) is responsible for enforcing the requirement instance assigned to it. This model also shows the structure of the interaction among agents in order to cooperate to achieve a higher level goal.

Fig. 5.1. Simplified View of OPTIMIS Objectives and Operations

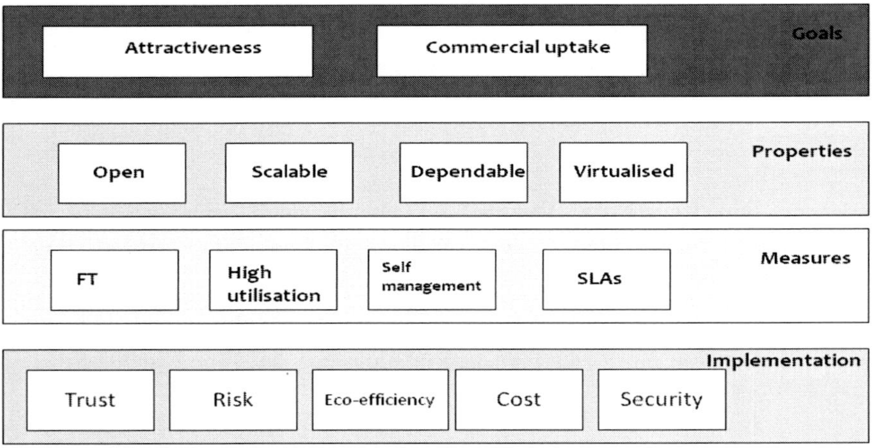

To reinforce the attractiveness and the commercial uptake of the cloud services, several issues need to be addressed such as openness, scalability, dependability, and virtualisation. For end-users, the operation of cloud services should occur transparently, i.e. without undue technical overheads. End-users must feel confident that the Quality of Service (QoS) they require will be met in the fulfillment of the contracts passed with the SP, which have the same TREC requirements regarding the contracts passed with IPs. SPs want to determine the best possible way to deploy their services. IPs, for their part, want to determine an optimal allocation of their cloud resources. To this end, they need an effective monitoring of the workload and a QoS based scheduling. Also, the design of suitable business models that define the relationships between the end-users, SPs, and IPs are necessary so that demand and offering of cloud services meet in the most efficient way.

These issues are supported by measures like the use of SLAs, self management, high utilisation of resources, and fault-tolerance mechanisms. The use of SLAs is a prerequisite and a building block for the wider adoption of cloud services. Self management, cost effectiveness, high utilisation, and fault-tolerance all refer to supporting processes that the SPs and IPs will set up in order to meet the cloud services demand in the most profitable way. The IP has to monitor its resources, to act before SLA violations, and to manage its infrastructure to provide QoS. Most of the time, the IP will favor the highest possible utilisation rate and if necessary use fault-tolerance mechanisms to ensure service operation.

The goal of the SP is to find the best possible IP to run the service (according to trust, risk, eco-efficiency, and cost goals) whereas the IP attempts to maximise its internal business goals (guided by revenue as well as other trust, risk, eco-efficiency, and cost goals).

The business requirements (primarily non-functional) need to be translated into and implemented as technical functions. These requirements come from interviews and conversations with a variety of stakeholders in the cloud industry, supplemented by additional primary research in the context of the work carried on business issues in clouds, and are grouped as follows (OPTIMIS 2010b; OPTIMIS 2011a): (a) flexibility and agility; (b) value added services, trust and control, and (c) cost efficiency. From the legal perspective, at the outset the requirements are split into three main aspects (OPTIMIS 2011b): (a) data protection and data security; (b) intellectual property rights, and; (c) green legislation issues.

2.2 Requirements from Use Cases

The studied use cases represent different scenarios in which OPTIMIS results are applicable, and have impact in previously identified roles. The main beneficiaries of the results of the project are the following. A "Cloud Programming" use case that will demonstrate the OPTIMIS Toolkit applicability for SPs both for the construction of new services but also to enable existing legacy applications to take advantage of the cloud deployment. A "Cloud Brokerage" scenario will demonstrate the OPTIMIS toolkit applicability for cloud Brokers or Aggregators as well as SPs. This use case focuses in the demonstration of the Multi-Cloud scenario. Finally, a "Cloud Bursting" use case will demonstrate the OPTIMIS toolkit applicability for final users, with the interaction between internal and external clouds, and for infrastructure providers that rely on third parties to offer their services. This use case focuses in the demonstration of the federated cloud scenario.

3. Architecture

The following description of the OPTIMIS Toolkit is focused around the challenges outlined in the introduction. The tools for self-assessment of non-functional criteria such as trust, risk, eco-efficiency, and cost are outlined first as these are used in all management decision making. To highlight the importance of the OPTIMIS Toolkit for service lifecycle optimisation, next follows a description of toolkit components for service construction, service deployment, and service operation, respectively. The last part of the section is an overview of the challenges associated with securing multiple connected clouds and the technologies applied to address them.

Tools for Self-assessment of Trust, Risk, Eco-Efficiency, and Cost

As discussed earlier, one of the overall objectives of the OPTIMIS Toolkit is to provide self-management capabilities for cloud infrastructures and services based on functional as well as non-functional criteria. The following describes how SPs and IPs assess aspects of TREC during service deployment and operation. An SP is responsible for matching the end-user requirements with the correct IPs to ensure the required demand is met. To achieve this, the SP service deployment process uses the TREC tools to rank IPs accordingly. On the IP side, TREC factors increase the performance and quality of an IP. For example, the estimated risk for an SLA violation supports the IP's decision of agreeing an SLA as well as initiating fault-tolerance mechanisms to prevent SLA violation (Djemame et al. 2011a). Thus, TREC tools improve the IP's reliability and QoS. In the following, the TREC tools are described.

Trust. The main function of the Trust Framework tool is to assess the trust associated with SPs and IPs. This trust assessment involves the following:

(1) **Main Trust**: this value is calculated using information from the monitoring and the SLA system of OPTIMIS plus an internal social network generated within the framework to manage the relationship between the actors
(2) **Social Network Trust**: which is calculated using the information stored in the social networking tool only
(3) **Self Experience Trust**: this value is calculated using the SP's experience only.

The trust framework includes the following components:

- **Social Networking Tools**: to interconnect all the actors making use of the Trust Framework. Thus, it creates links between end-users, SPs and IPs with an associated satisfaction rate. This rated interconnection gives the possibility to assess the trust in relation to one of the actors based on the opinion of the end-users who have interacted in the past with the SP or IP;
- **Trust Framework Core**: supports the business logic of the Trust Framework, and is comprised of: (a) Data Asset: which stores the information obtained from the monitoring system and the social networking tools, as well as the historical information about the actor ranked, and (b) Trust Engine: where all the logic to generate the reputation of the actors is located
- **Monitoring**: a set of monitors that collect data from different sources, for example performance of a service on the SP side, SLA monitoring, scalability rules on the IP side.

Risk. To support the risk assessment functionality on the SP side, a number of components are identified: a Confidence Service (comprising a Risk Assessor and a Provider Assessor), a Risk Inventory and a Historical Database for recording past SLA transactions. The confidence service will take into account the various risks of working with the different IPs accessing the providers. This is used in the service deployment process to make decisions based on a stored database of history of working with the different IPs and the risk inventory associated with the different assets involved. The Risk Assessor provides the functionality to evaluate the risk associated with an offer returned by the IP's Admission Controller (AC), by determining the reliability of the offered Probability of Failure value from the IP's SLA quote (Djemame et al. 2011b). The Provider Assessor verifies the expected integrity of an IP's guarantees when they make any SLA offer through an implementation of Dempster-Shafer Analytical Hierarchy Process (DS-AHP), whereby each decision alternative (in this case each provider) is mapped onto a belief and plausibility interval (Huan et al. 2008). A risk inventory is a simple database of risks associated with each asset, their vulnerabilities and threats. This also contains risk mitigation strategies following risk assessment.

Performing risk assessment at the IP level increases the performance and quality of the IP. Thus, the IP also assesses the service to be deployed by determining an estimated risk if it were to accept the SLA taking into account fault tolerance mechanisms and actions following an SLA violation, in turn improving the IP's reliability and QoS. Four categories of risk are identified: technical, policy, general, and legal.

The IP would forward the "service manifest" request to the AC to evaluate the feasibility of admitting the new service, with respect to current infrastructure load, predicted future capacity, as well as risk. This helps the IP to determine where to place the Virtual Machines (VMs) by combining its local management policy with the function and non-functional requirements. The Cloud Optimiser (CO) also supports risk assessment at the service operation to ensure the SLA fulfillment. In addition to a risk assessor, a risk inventory, and a historical database, a consultant service component is identified at the IP side, which uses data mining tools on the history of events of running similar services or working with the same SP to estimate risk. The consultant service can also have access to all the monitoring information, keeping the IP on track with the changes. This data can be static or dynamic in nature regarding its resources and the current service execution. Examples of such information are the current workload, system outages, temporary performance shortages, monitored network traffic, experts' availability, or general information regarding the number of services to operate. The monitored data helps to determine bottlenecks in the IP's infrastructure so that the provider can improve its capacity planning, administration, and management of its resources. This leads to high, cost-effective productivity of virtualised resources. Some of the risk models are probabilistic, possibilistic, or hybrid.

Eco-Efficiency. This tool is in charge of assessing energy-related aspects, such as energy efficiency and carbon emission levels, in a given cloud infrastructure. In this sense, it uses the information collected by the respective monitoring tool of the basic toolkit. This assessment is fully guided by ecological-aware rules defined according to the energy-related parameters, for example carbon emission levels, specified in service manifest. Once this assessment is performed, it is used by self-management policies that act in accordance to provider's energy requirements and BLOs. Between the possible management actions we find the consolidation and migration of VMs where services run. The tool is composed of the following components:

(1) **Eco-efficiency assessor**: the core sub-component and responsible for performing the aforementioned assessments and predictions. The assessments in this component result in action requests to be performed by the VM Manager, such as requests for migrating a VM, or consolidating several VMs in a single node

(2) **Energy efficiency monitoring**: contains the information collected by the respective monitoring tool and is used by the assessor

(3) **Eco-efficiency rules**: where rules are defined according to the energy-related parameters specified in services' manifest, and which fully guide the assessments.

Cost. This tool is responsible for assessing and predicting the relevant costs for service deployment and operation from both the SP and IP perspectives. The module can be deployed independently by both the SP and IP, and is used by these actors to evaluate and predict their costs and subsequently to optimise their profitability. The tool is composed of three main components:

(1) **Cost Assessor**: responsible for managing all operations within the cost tool, and is used to accurately trace the absolute costs of service operation and deployment. In addition the Cost Assessor acts as a prediction engine to assess various scenarios, for example how much would it cost to host a service with specific SLA requirements e.g. CPU, Disk, Memory etc. The cost assessor may also recommend the most cost effective policy to employ when required to violate SLAs.

(2) **Cost Modeller**: responsible for defining the mathematical models to accurately record and predict the cost of service deployment and/or operation. To build these models the modeller utilises statistical techniques such as data mining on the data stored within the cost database. For the SP this may include data regarding hosting charges (from IP), service usage, SLA levels etc. For the IP this may include hosting income (from SP), service usage, SLA levels, infrastructure utilisation, energy usage etc.

(3) **Cost Database**: a persistent store for all previous cost analyses (predicted/realised) and relevant cost data which has been gathered from the monitoring and SLA/SLO managers. This is subsequently used for the definition and refinement of the cost models.

4. Service Construction

This section describes the different aspects of defining a service with for an OPTIMIS IP. We give an overview of the service manifest, which describes a service to be deployed. Moreover, we discuss how software licenses are managed in the OPTIMIS environment.

4.1 Service Manifest

The OPTIMIS manifest is the global description of a service that is delivered by an OPTIMIS IP. It is generated by the OPTIMIS SP during the service construction phase and consumed by the different IP components during the service negotiation and instantiation phases. The manifest basically consists of four sections: (a) the service description section; (b) the TREC sections; (c) the elasticity section; and (d) the data protection section. Figure 5.2 below outlines the service manifest structure.

Fig. 5.2. Overview of the service manifest structure

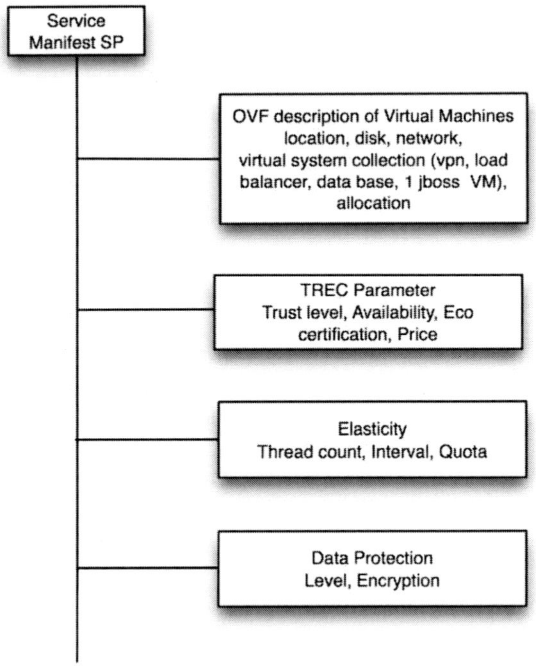

The service description section is a functional description of the cloud service to be provided. It consists of two parts, an abstract service description and an allocation section. Since the infrastructure resources are provided in form of VMs, the Open Virtualisation Format (OVF) is used for describing these. The OVF document defines a collection of virtual systems, where each virtual system corresponds to one service component and is uniquely identified by its OVF i.d. In addition, the allocation section defines a set of allocation constraints such as minimum and maximum instances of each virtual system (service component). The service description section is complemented by a TREC section. The TREC section defines non-functional parameters of the service, i.e. requirements related to the energy efficiency, reliability, and availability of the allocated resources. Within the elasticity section of the service manifest the SP can specify a set of elasticity rules, i.e. in order to start up a new VM when a defined system load is reached, or to shut down a virtual system instance when system load falls below a given threshold.. Finally, the SP can specify constraints related to data placement in the data protection section, for example it is possible for the SP to specify a list of countries where service data may (or may not) be placed.

4.2 License Management

Access to licenses for authorising the execution of an application in a cloud located beyond the administrative domain of the site running the license server, usually leads to applications aborting during startup because of unreachable license servers, for example due to firewall issues or a network address mismatch. In OPTIMIS, these problems are addressed with a prototype for licensing and license management developed in the European project SmartLM (Cacciari et al. 2010), which provides licensing technology for location independent application execution. Through the decoupling of local authorisation for license usage from authorisation for remote application execution, SmartLM provides the necessary mechanism for transparent enabling license-protected applications in a cloud environment.

Software tokens are created when usage of a particular software license is granted to a user upon request by the local license management system. The tokens carry the authorisation information based on the user's request, locally defined policies, and availability of a license or feature for running an application. As software artifacts, the tokens are mobile, thus providing the necessary flexibility for licenses to follow applications into clouds.

At deployment, the VM Contextualiser retrieves and embeds license tokens into the VM hosting the application. No communication between the application and the license server that issued the token is required at runtime.

In the future, OPTIMIS will extend the SmartLM solution by providing: (a) dynamic deployment of a trusted instance managing a number of tokens for one or more applications, and (b) dynamic deployment of a full license service with a subset of the licenses available in the users' home organisation. The configuration of the dynamically deployed license service will be managed by the VM Contextualiser. This approach is especially useful when the same cloud is used to run licence protected applications for a long time. In the first approach, the contextualisation tools are responsible for configuring and deploying the trusted instance for the respective network environment and to transfer tokens.

5. Service Deployment

After service construction, deployment of the service to an IP is assisted by the "Service Deployer" (SD). The SD offers a unified interface for service deployment and management, encapsulating the internal component interactions triggered as a part of deployment. This process is illustrated in figure 5.3 below, and the following list presents the general steps of service deployment from the perspective of the SD, although the exact interactions may vary slightly between different deployment scenarios (multicloud, bursting, etc.).

(1) Prior to deployment, the SD is configured with properties relevant to the different scenarios, for example the endpoints of several IPs in the multi-cloud scenario. The deployment process is initiated by passing the service manifest to the SD.
(2) The SD depends on an internal component, the "Service Deployment Optimiser" (SDO), to rank existing deployment alternatives according to the needs of the service. The SDO negotiates with QoS components of different IPs, and ranks the alternatives based on, for example, the TREC factors.
(3) When a suitable IP for deployment has been found, the SD triggers "contextualisation" (configuration) of the VMs, and uploads the VM images to the repository of the chosen IP.
(4) Once the data transfer is completed, execution of the service is initiated by calling the local "Service Manager" component, which is responsible for managing the service in the operation phase.

Each general step outlined above triggers a series of underlying interactions not visible to the SD. For instance, in the IP, the QoS component relies on an Admission Control component in the negotiation process, encapsulating the potentially complex statistical models and state management of the infrastructure. The following sections describe the inner workings of the major supportive components in more detail.

Fig. 5.3. The general steps of service deployment. The deployment is initiated with the Service Manifest (1). A suitable placement is found by contacting the SDO (2). The raw VM image is contextualised with the necessary information (3), before the service is finally deployed (4)

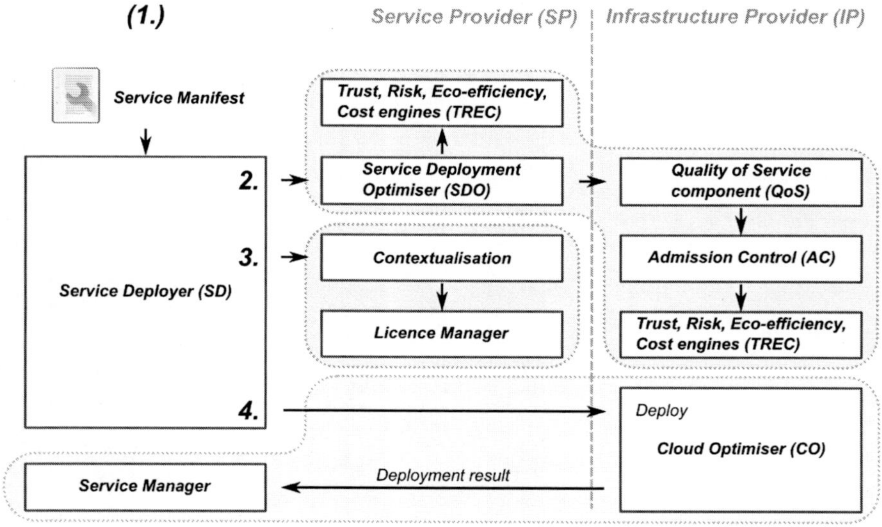

5.1 Quality of Service

Current cloud environments are offered to their customers in a best effort approach. Instead of guarantees, a statistical uptime expectation is communicated to the user with minimal compensations in case of unexpected downtime. In contrast, in OPTIMIS, SPs can rely on SLAs with IPs, for example, to extend their own resources dynamically with cloud resources in case of peak demands of their customers. These SLAs cover aspects like TREC factors, security, legal requirements for data-placement, etc.

The OPTIMIS QoS SLAs are based on the WS-Agreement for Java (WSAG4J) implementation (see figure 5.4). WSAG4J (http://packcs-e0.scai.fraunhofer.de/wsag4j) is a full implementation of the WS-Agreement standard of the Open Grid Forum (Andrieux et al 2007); it provides comprehensive support for common SLA management tasks such as SLA template management, SLA negotiation and creation, as well as SLA monitoring and accounting. The OPTIMIS QoS component interacts with the following three components of the OPTIMIS toolkit:

Fig.5.4. WSAG4J components

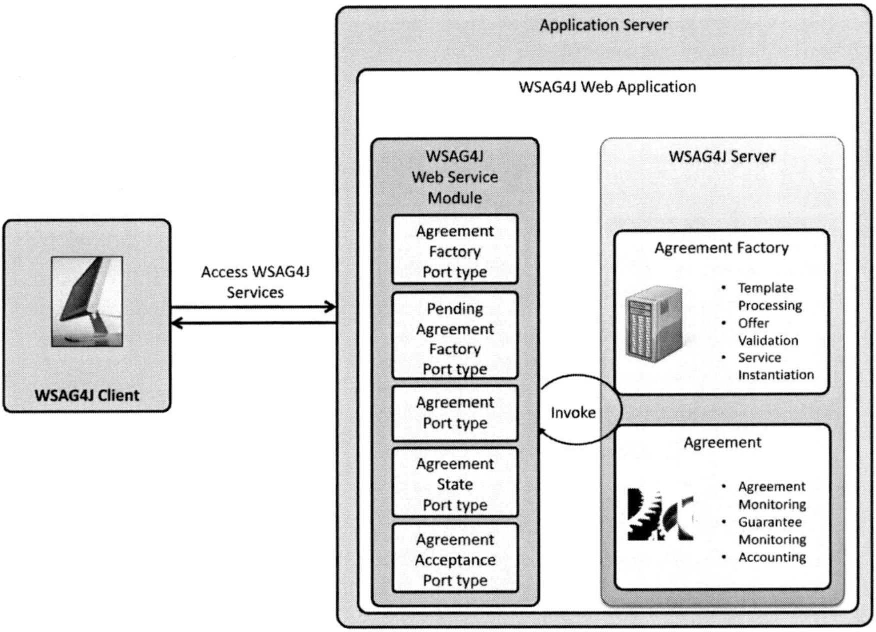

- **Admission Control**: In order to negotiate and create SLAs with a service consumer (the OPTIMIS SP), the service provider (the OPTIMIS IP) must assert that the agreed service can be provided. When a SLA is created with the service consumer, the SLA management component of the service provider initiates the service deployment process.
- **Service Monitoring**: Each provider that offers SLAs for a particular service should perform SLA monitoring in order to detect possible SLA violations and to take appropriate counter measures. The SLA monitoring component therefore integrates with the Service Monitoring, which provides the required monitoring data to assess the fulfillment of the service level objectives defined in a SLA. SLA monitoring is a continuous process during the lifetime of an SLA and is performed on a regular basis. The SLA guarantee evaluation results in corresponding events, which are then processed by other components, such as SLA accounting (penalties and rewards), the TREC toolkit, or the Cloud Optimiser.

- **Cloud Optimiser**: The Cloud Optimiser (CO) consumes events generated by the SLA monitoring of the IP. According to the event type and its importance, the CO takes appropriate measures in order to fulfill the SLA guarantees. These measures may include starting up or shutting down a VM upon events from the Elasticity Engine, or to migrate or restart a VM.

5.2 Admission Control

Admission Control (AC) in OPTIMIS has the main goal of deciding whether a new service can be admitted by the IP, and deciding what the optimum allocation of the VMs of the service to the available underlying resources is, in case of acceptance. This task is far from trivial, since in many cases, admitting a new service increases the risk of already deployed ones failing. Conversely, a strict acceptance policy results in increased rejections, reduced utilisation, and thus lost revenue.

Hence, in order to come to a decision, AC must weigh up critical factors such as the current workload of the infrastructure, the requirements (in terms of computing, memory, storage and networking resources) of the new service, and the IP's business policies in terms of risk, trust, eco-efficiency, and cost. For example, in some cases it may be more lucrative to admit a highly profitable service even under the risk of jeopardising existing, but less profitable, ones. Likewise, given a very sensitive IP in terms of eco-efficiency and two services competing for the same resources, the less profitable service may be accepted if it is sufficiently more eco-efficient.

In the context of OPTIMIS, an IP is a business entity that establishes SLAs with SPs to allow for booking a set of virtualised resources for hosting distributed services, whose components are encapsulated inside VMs. According to the expected usage of the service, the SP specifies a lower and upper limit for the number of VMs (called basic and elastic VMs respectably) that may be required during runtime for each service component. At the first activation of the service, only the basic VMs participate in the service. During runtime, elastic VMs may be added to the service according to the workload and the SP's policy. This number of additional elastic VMs cannot exceed the upper limit defined in the SLA.

In finding the optimal allocation, AC makes use of an optimisation model that builds upon factors such as the performance requirements of the services as defined in the SLA, and information about the current workload status of the infrastructure's resources acquired by communicating with OPTIMIS Monitoring system. Furthermore, the AC also accounts for business rules, expressed in TREC terms provided to the AC by the IP. These factors are mathematically modelled in terms of constraints and an objective function to be optimised.

In addition, differently from the traditional admission control mechanisms, the OPTIMIS AC builds on a probabilistic basis. This is due to the fact that the number of elastic VMs may be quite large compared to the basic requirements for the service. For example, given a service with a high variation in the number of users, the number of elastic VMs that may be required to be added at runtime can be many times more than the basic VMs. Therefore, the number of elastic VMs plays a significant role in the total requirements and the cost for hosting a service. Thus, the IP has a strong interest in investigating the possibility of narrowing the resources that need to be booked for elasticity reasons when accepting a service. At the same time, such an approach may increase the possibility of deviating from the agreed QoS level, i.e. the required availability of the resources, and the imposed penalties may also outweigh the advantages of this approach.

With dynamic elasticity of the number of VMs of a service in focus, OPTIMIS AC tackles the aforementioned problem of optimum allocation of such services on virtualised physical resources, by incorporating a probabilistic approach in terms of availability guarantees (i.e. minimum probability of finding the required number of VMs available when actually needed). The proposed approach relies on the actual probabilities of requiring extra elastic VMs for the services, which are incorporated into the admission control test, allowing a reduction of the number of physical resources that need to be booked for elasticity reasons. The output of the optimisation problem is the number of accepted elastic VMs per service component as well as the location (host) of each VM that comprise the service (both basic and elastic ones). The resulting optimisation problem constitutes an admission control test that may be run by the IP. Optimal allocation of distributed services under QoS constraints across multiple resources is an NP-hard problem. For modeling and solving this problem, the AC uses the General Algebraic Modeling System (http://www.gams.com/).

5.3 Contextualisation

The VM Contextualiser uses data gathered by the SD to set the context of a VM at runtime. This involves the SD assembling data from several sources and providing it as input to the public interface of the VM Contextualiser along with the ID of the base image to contextualise. The Data Manager is used to propagate VM images to the VM Contextualiser. Once the VM image has been propagated, it is mounted and modified to include scripts that interact with the guest operating system at boot time, preparing it to receive context in a reusable fashion. After the image has been prepared, it is returned to the VM Image Repository for storage and used later by the

VM Manager. The lightweight scripts embedded in a VM image bootstrap to additional scripts held within an ISO CD ROM image, as per the OVF recommendation (http://www.dmtf.org/standards/ovf), mounted within the guest operating system. At run time the bootstrapped scripts access contextualisation data stored within the ISO CD ROM image, giving a VM an identity. The inclusion of the ISO image as a mechanism to store contextualisation information provides a facility to separate the contextualisation data from the VM image. This removes the need to create unique VM images for each VM to be contextualised, while also improving the security of the contextualisation process as security certificates and keys are not stored in the VM image itself, but rather in the dynamically generated ISO image.

The VM Contextualiser uses QEMU (www.qemu.org), a generic and open source machine emulator and virtualiser, to manipulate images. The QEMU tool "qemu-img" enables the conversion of VM images. In addition, Linux system tools (such as "mount," "iosetup" and "kpartx") are used to mount VM images as loop devices for write manipulation. For the creation of ISO images, the Linux system tool "mkisofs" is used in addition to the previously mentioned tools to create and modify ISO images. The VM Contextualiser interacts with other components in the OPTIMIS Toolkit at the SP level during image-level contextualisation. The SD initialises the preparation of a VM image for receiving context data at runtime. When the VM Contextualiser receives a request to prepare a VM image, it retrieves the image from a VM image repository. After manipulating the image, it is sent back to the repository for safekeeping.

Figure 5.5 illustrates the instance-level contextualisation process of a VM at the beginning of its execution. During the boot sequence of a VM, the contextualisation tools mount an ISO CD image that contains contextualisation data and scripts to process this data into a useable form. These scripts communicate through a known interface to the OS specific contextualisation scripts embedded in the VM image. These OS specific scripts manipulate configuration files of associated OPTIMIS components and software dependencies that support the service, setting their context. If a component or software dependencies requires continuous update to its context, the scripts can remain in a daemon-like mode.

Fig. 5.5. Interaction between VM image and ISO image at run time

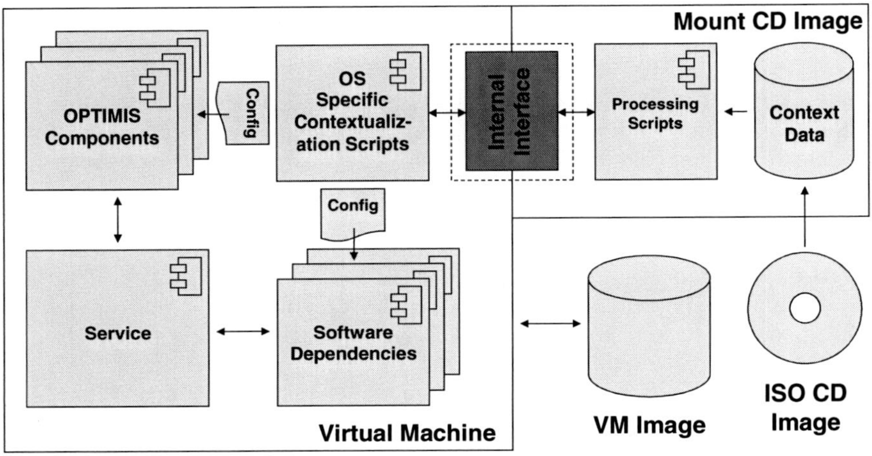

6. Service operation

An OPTIMIS IP makes use of several internal management procedures during service operation to supervise and fine-tune the execution of the service and optimise the utilisation of the infrastructure. The relation between different components is described in figure 5.6, and the following sections describe some of these internal procedures in more detail.

Fig. 5.6. The CO finalises service deployment and is the central component for service operation. Monitoring data is consumed by the FTE and EE, and may result in a change in the number of allocated VMs per service, which is managed by the CO with the support of the VM Manager and Data Manager components

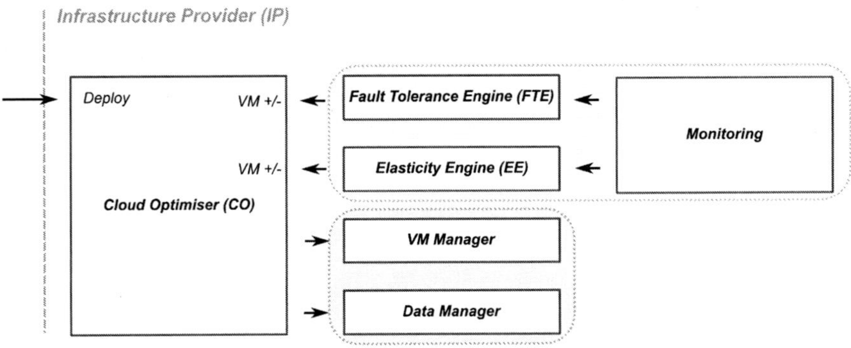

6.1 Monitoring

Deployment, execution, and management of applications in highly dynamic infrastructures such as clouds introduce a whole new set of requirements in respect of monitoring that must be addressed by developers and providers of the related services. The introduction of the virtualisation layer along with the energy efficiency directives applied into the data center operation field, have increased the amount of data that must be collected and processed (i.e. further parameters must be measured and monitored, such as the amount of memory used by a concrete VM or the CO2 emission of a data center, for example. Additional parameters are discussed in Katsaros et al. [2011]). Furthermore, the elastic character of the infrastructure, essential in a cloud environment, requires that the monitoring tool is able to handle a service or infrastructure scaling up or down dynamically. Apart from scalability, such a system should adopt the service-oriented design pattern, which is the keystone of the cloud computing paradigm. The existence of multiple layers (physical infrastructure, virtual environment, application), each one ultimately associated and dependent on the other, results in the need for collecting and aggregating all information towards an effective decision taking mechanism. Furthermore, the storage of this data must be performed in an efficient way so that they can be reused by other components of the OPTIMIS Toolkit, such as performance estimation mechanisms.

As far as the monitoring infrastructure concerns, the proposed architectural model incorporates the conceptual *Information Provider* layer (figure 5.7 below). By this we mean the different sources from where monitoring data is being collected. As Information Providers, we consider the main entities producing information: the physical infrastructure, the virtual infrastructure, the application specific metrics, and the energy efficiency measurements. The actual collection of information is performed by respective components known as "Collectors." We have designed our solution in such a way that this layer is extendable, allowing the incorporation of additional sources through corresponding collectors. This introduces a first degree of scalability (in the sense of extendibility), which enables the monitoring system to handle a growing number and type of monitoring parameters. Data collection is realised either with a push model where the Collectors gather data and pushes a monitoring report towards "Aggregator" components in the higher conceptual layer, a corresponding pull model, or a combination of the two. More details about the operation and the implementation of the monitoring collection mechanism can be found in Katsaros et al. (2011).

Fig. 5.7. Monitoring infrastructure overview

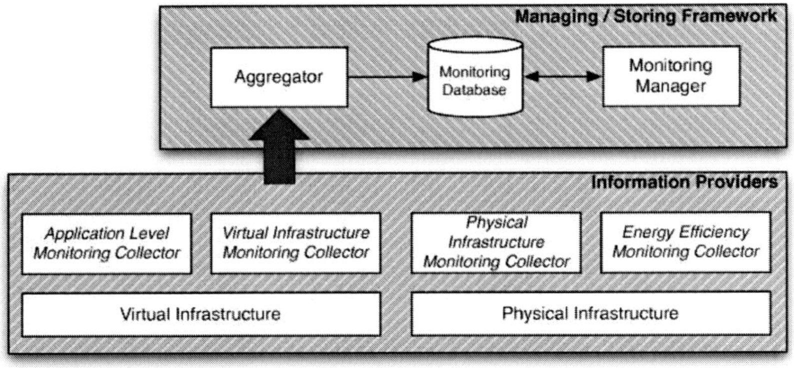

The second hierarchical level of the proposed approach is the "Managing and Storing Framework." This layer includes the components in charge of managing the data collected from the monitoring information providers and storing them in an aggregated way in a monitoring database. The aggregation is done in terms of data model consistency, which is designed in a way that eases both the process of storing the information and the subsequent retrieval of this data for analysis and usage. Each monitoring report must thus contain an identifier that relates it to the source of information. This is based on the concept of entities (IP, SP, service manifest, service) and the relationship between them. This contributes to a second level of scalability. As presented in Katsaros et al. (2011), the collected data is classified and aggregated according to those entities, which results in a number of database records, following the schema of the designed data model. The structure of the database conforms to best-practices for database scalability, but a discussion of these is out of the scope for the work presented here.

The "Monitoring Manager" component serves as the orchestrator of the whole monitoring process. The role of this component is twofold and includes the capabilities of controlling the monitoring process (start/stop actions) and providing the necessary interfaces towards other components of the cloud ecosystem as well as to other external consumers of monitoring information (e.g. graphical user interfaces, administrators etc.). The collected and aggregated data can be provided in two different ways: on the one hand, the most recent monitoring information is provided in an on-the-fly manner and, on the other, the monitoring data is being stored in a local database with historical information of the deployments and executions. Notably, the monitoring components do not provide event processing capabilities themselves, but the Monitoring Manager provides the necessary interfaces for the TREC self-assessment tools to acquire and analyze the required data and generate the corresponding events, if needed.

A third level of scalability is introduced by the general architecture design and the concept of the OPTIMIS Toolkit. Through this, some of the toolkit components are decentralised, being part of the SP domain, while others are located in the IPs Ferrer et al. (2011).

6.2 Elasticity Engine

One of the most prominent features of cloud computing is the ability to quickly increase or decrease the amount of resources allocated to a service, either by modifying the allocation to each running VM ("vertical elasticity") or by modifying the number of copies of VMs running on the infrastructure ("horizontal elasticity").

Several different models and approaches can be applied to elasticity. Ali-Eldin et al. (2011) present a queuing theory based approach with different combinations of reactive and proactive models for scaling up or down services. Gambi et al. (2010) employ mathematical surrogate models to protect SLAs though the means of elasticity. Elasticity support in the OPTIMIS framework is designed to be agnostic to the actual elasticity decision algorithm used, creating a software base upon which several different models can be used. The "Elasticity Engine" (EE) component contains a framework for interaction with supportive components, and an easy to modify module containing the actual model for elasticity. The primary interactions of the EE are with the CO (to start or stop VMs) and the monitoring system (to gather runtime status) and these interactions are isolated from the internal model for elasticity. The EE is also designed to be a stand-alone tool usable on the SP side, in which case the SP itself replaces the interactions between the EE and the CO. Even though the below description focuses more on IP-driven elasticity, the process in the SP side is similar.

The elasticity process is initiated by the CO on a per-service basis, supplying the EE with a set of "elasticity profiles" (or "elasticity rules" [Rodero-Merino et al. 2010]). The rules specifies, for example, the values, or "Key Performance Indicators" (KPIs) to monitor along with the relation between the KPI and the amount of VM resources (either size of each VM or number of instances) granted to the service. Updating the elasticity rules of a service is one way of adapting the service to different objectives. Note that the exact interpretation of elasticity rules and the manner in which they are enacted is different depending on the internal model used, but the different interpretations do not affect the general management procedure.

Once the process is initiated, the EE collects data on the relevant KPIs by communicating with the Monitoring system. The received data is an input to the internal model in the EE, which returns an estimation of the required amount of resources for this particular service. The estimation is passed along to the CO, which is responsible for the process of actually adjusting the amount of resources dedicated to the service. Notably, the EE requests may be ignored (likely resulting in SLA violations) if the CO policy deems this to be more aligned with the overall management goals of the IP.

6.3 Cloud Optimiser

The CO combines the Monitoring system and the fundamental TREC assessment tools with a set of management engines to perform cloud infrastructure self-management governed by BLOs of the IP. Notably, goals in these higher-level objectives (policies) typically differ from those in the SLAs established with SPs, thus self-management decisions become tradeoffs between the BLOs and SLAs.

The CO is also responsible for tracking the relationship between services and the associated set of VMs where they are running. The CO updates this matching between services and VMs to the Service Manager when a new service is deployed, a VM is added or removed (due to elasticity), or a VM is restarted upon a fault. Upon deployment of a new service, the CO notifies the EE and Fault Tolerance Engine (FTE), the latter a self-management component responsible for ensuring robust VM provisioning.

In addition to communication with these engines, the CO interfaces with the VM Manager (based on Emotive Cloud [www.emotivecloud.net]) and Data Manager, two high-level actuators for handling of VMs and Data, respectively. The Data Manager includes, in addition to data transfer capabilities, a framework for handling service data protection issues, for example performing storage location selection based on legal deployment constraints. The CO also contains a policy repository that enables IP administrators to change or reconfigure policies for how fault tolerance, admission control, VM management, and data placement should be performed. In summary, these policies in the CO determine the most suitable management actions of the OPTIMIS components in order to meet the IP's BLOs.

6.4 Service Manager

The Service Manager (SM) component serves as a management point for services during deployment and operation. In essence, it keeps the necessary information about a service: details about the selected IP where the service is being deployed, how many and which VMs it uses as well as the Internet Protocol addresses of the VMs. The SM component is a part of the SP's tools of the cloud framework but it also provides external interfaces towards the IP level. The SM interacts with the following components:

Service Deployer (SD): during deployment a set of service state is created within the SM for every service, including all the above-mentioned information. After the SD acquires the initial set of information during the service negotiation process, it triggers creation of service state data in the SM.

Elasticity Engine (when used at SP level): the SM initiates the EE for a service each time a new service resource is created. Thereafter, the EE evaluates the data received by the monitoring infrastructure and performs elasticity actions accordingly.

Cloud Optimiser (IP level): in case a service management action is taken (for example additional VMs are added for a service), the CO notifies the SM of the related change on the service deployment. Thereafter, the adequate service state data is updated.

7. Inter-Cloud Security Framework

Of the various high-level concerns being addressed by the OPTIMIS project, the provisioning of a secure communication framework to the services utilising the resources of different cloud IPs is of high importance. The usage pattern of these services should be quite flexible, i.e. on one hand they might be directly accessed by end-users or on the other they might be orchestrated by other SPs for their customers. The presence of the multiple IPs in the cloud computing ecosystem is a key issue that needs to be addressed by any inter-cloud security solution. In addition to the IPs, the other three main actors in the cloud ecosystem are SPs, end-users, and cloud brokers. For the numerous interaction possibilities among these actors, whatever the usage scenarios may be, the security of data and communication between the service consumers and its multiple providers is of paramount importance.

In light of the previous discussion, it is clear that an inter-cloud security solution providing a framework enabling seamless and secure communication between the actors of a cloud ecosystem over multiple cloud platforms is highly desirable. Such a solution, however, has a number of associated challenges because of architectural limitations. Most current cloud service platforms and the multi-tenants environments they offer make it difficult to give consumers flexible and scalable control over the core security aspects of their services like encryption, communication isolation and key management, etc. Secure communication is also challenged by lack of dynamic network configurability in most cloud providers, caused by the inherent limitations of the fixed network architectures offered by these providers. We address the secure, flexible and scalable communication concerns that in our view must be overcome in order to provide holistic cloud services to consumers from multiple cloud service providers. We present the architecture and design of an inter-cloud secure communication framework that offers the features of dynamic and scalable virtual network formation, efficient and scalable key management, and optimised peer discovery, on top of secure and private communication between consumers across multiple cloud platforms.

Our solution provides a single virtual network to the consumers of cloud services using the infrastructure and resources from multiple cloud providers and offers all the management capabilities to efficiently and transparently run services on top of this network while catering for the dynamic growth and shrinkage of the network and its participant entities. At its core, it provides the ability to automatically establish peer-to-peer overlay networks comprising the VMs and other infrastructure resources constituting a cloud service. Using the same P2P techniques, we also offer a distributed key management service which facilitates the automatic discovery of the peers participating in a service and the binding of cryptographic constructs like keys, certificates and fingerprints to their identities. Several key management models exist currently that can be utilised for our purposes, ranging from centralised key distribution centres to those based on multicast communication (Qiu et al. 2006) or distributed hash tables (Rafaeli and Hutchinson 2003). The Inter-Cloud VPN (ICVPN) architecture consists of the following main components.

Peer-to-Peer Overlay. The core technique employed by the ICVPN is the secure tunneling of network traffic over a scalable, resilient and self-managing peer-to-peer overlay with minimal manual configuration and administration. This approach acts as an aggregation service for the peered resources (which in this case are VMs) spanning multiple cloud domains to help form a virtual private network. The overlay can be further leveraged to customise the virtual topology of the network, by either emulating a completely distributed mesh topology at one end, or a centralised star topology by automatically electing a master peer to act as the gateway of the virtual network at the other, or an in-between hybrid topology using a number of selected super-peer nodes. The adaptation of a suitable network topology is indeed a major communication benefit for services using P2P overlays for their operational requirements (Dinger and Hartenstein 2005). Figure 5.7 shows the presence of all the above mentioned topologies in the three cloud infrastructure providers.

Fig. 5.7. Establishment of a peer-to-peer overlay based virtual network on top of multiple cloud infrastructure providers

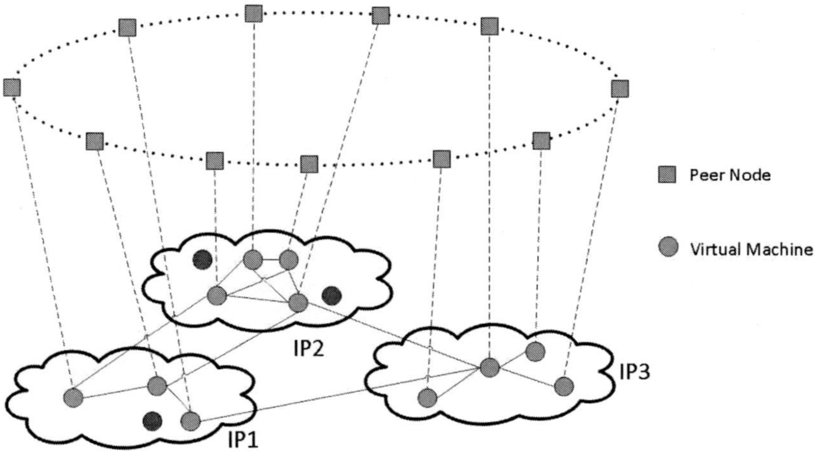

In our use case scenario, the end-users or the SPs are responsible for provisioning VMs from cloud providers to deploy and run their services. These VMs can be considered as the peers of the overlay network if they want to join the virtual private network and the addition of a peer to the overlay network is handled by a P2P client embedded in the virtual appliance image used to instantiate a VM on a cloud platform. Upon joining an overlay network to become part of a service, a peer starts the process of creating secure tunneled connections to the other peers of the service according to the network topology selected for the operation of that particular service. The typical constructs used for this purpose are TAP and TUN devices, which are virtual network kernel devices set in the operating system of the peer node (http://vtun.sourceforge.net/). A TAP device simulates an Ethernet interface card and operates with link layer datagrams such as Ethernet frames. A TUN device simulates a network layer device and operates with network layer packets such as IP packets. TAP is mainly used to create a virtual network bridge, while TUN is used with routing, so choice of their selection mostly depends upon the network topology chosen for the overlay network (http://en.wikipedia.org/wiki/TUN/TAP).

Secure Virtual Private Connections. The key feature of the ICVPN is the establishment of a secure communication network between the peers of the overlay formed over a collection of cloud providers' infrastructure. To achieve this, we make use of public key cryptography which supports the Public Key Infrastructure (PKI) model. This model makes it possible to create authenticated and confidential end-to-end tunnels that provide protection against eavesdropping, message tempering and message forgeries. Another practical advantage of using PKI is the reuse of existing frameworks and tools which have been thoroughly tried and tested in a myriad of different domains, are widely used, and have been adopted and integrated in countless security techniques.

The P2P client software sets up and configures the TAP/TUN device automatically and assigns unique IP addresses to their virtual interfaces and DNS hostnames to the peers themselves. They can also be used as peer IDs and this record is put on the Distributed Hash Table (DHT) for the operation of the peer discovery service of the overlay network, which works by obtaining the list of unique peer identifiers and mapping them to the locally assigned IP addresses and host names. The same DHT is used for binding security credentials like certificates and keys to the unique peers and identifies which service the peers are a part of. Once a peer has obtained the X.509 certificate of a peer it wants to communicate with from the DHT, they can directly negotiate and generate symmetric session keys for encrypted tunneled communication. These certificates are signed by a Certificate Authority (CA) trusted by the end users of the service or the SP. The CA model could be replaced by a cloud-oriented trust model, such as the one in the OPTIMIS toolkit.

Once equipped with the required security credentials and peer locations, the peers can proceed to form secure tunnels to complete the VPN construction. The DHT is used as the control channel that provides the meta-data to carry out the necessary configurations and connection setups.

8. Concluding remarks

With the five identified challenges for future clouds as a point of departure, our diverse research activities range from control theory for elasticity to cloud business models. Our research findings are incorporated into the OPTIMIS Toolkit, a set of self-management tools for cloud service and infrastructure optimisation that can be used to implement clouds of multiple possible architectures. All toolkit management actions are harmonised by policies that incorporate aspects of trust, risk, eco-efficiency, and cost, without neglecting performance. Future directions for our work includes, for example, research on policy-driven service management and investigations of migration and service relocation for continuous service lifecycle optimisation.

Acknowledgments

We thank all members of the OPTIMIS consortia for their contributions to this work. We also acknowledge the anonymous reviewers for their valuable feedback. Financial support has been provided in part by the European Commission's seventh framework programme under grant agreement no. 257115.

References

Ali-Eldin A., Tordsson J. and Elmroth E.. "An Adaptive Hybrid Elasticity Controller for Cloud Infrastructures". Submitted. 2011.

Andrieux A., Czajkowski K., Dan A., Keahey K., Ludwig H., Nakata T., Pruyne J., Rofrano J., Tuecke S. and Xu M. "Web Services Agreement Specification (WS-Agreement)". GWD-R (recommendation), Open Grid Forum, GFD.107 recommendation, 2007. http://www.ogf.org/documents/GFD.107.pdf (accessed October 12, 2011)

Cacciari C., Mallmann D., Zsigri C., D'Andria F., Hagemeier B., García Peréz D., Rumpl A., Ziegler W., Gozalo M. and Martrat J. "Software Licenses as Mobile Objects in Distributed Computing Environments." Euro-Par 2010 Parallel Processing Workshops, LNCS 6586, 279–286, 2010.

Dinger J. and Hartenstein H. "On the Challenge of Assessing Overlay Topology Adaptation Mechanisms," IEEE International Conference on Peer-to-Peer Computing (P2P'05), 145–147, 2005.

Djemame K., Armstrong D., Kiran M. and Jiang M. "A Risk Assessment Framework and Software Toolkit for Cloud Service Ecosystems." To appear in the proceedings of the Second International Conference on Cloud Computing, GRIDs, and Virtualization, Rome, Italy, September 2011 (2011a).

Djemame K., Padgett J., Gourlay I. and Armstrong D. "Brokering of Risk-Aware Service Level Agreements in Grids." *Concurrency and Computation: Practice and Experience* 23 (7) (2011b).

Ferrer A.J., Hernández F., Tordsson J., Elmroth E., Ali-Eldin A., Zsigri C., Sirvent R., Guitart J., Badia R.M., Djemame K., Ziegler W., Dimitrakos T., Nair S.K., Kousiouris G., Konstanteli K., Varvarigou T., Hudzia B., Kipp A., Wesner S., Corrales M., Forgo N. , Sharif T. and Sheridan C. "OPTIMIS: a Holistic Approach to Cloud Service Provisioning." *Future Generation Computer Systems* 28(1) 2012: 66-77.

Gambi A., Toffetti G. and Pezze M. "Protecting SLAs with Surrogate Models." ACM, 2010.

Huan, Z. Gong B. and Xu X. "A DS-AHP Approach for Multi-attribute Decision Making Problem with Incomplete Information." *Journal of Expert Systems with Applications* 34 (2008): 2221–2227.

Katsaros G., Gallizo G., Kübert R., Wang T., Fito J. O. and Henriksson D. "A Multi-level Architecture for Collecting and Managing Monitoring Information in Cloud Environments." International Conference on Cloud Computing and Services Science (CLOSER), 2011.

OPTIMIS Project. "Requirements Analysis." Deliverable 1.1.1.1, September 2010a.

—. "Cloud Business: Needs, Desires, Concerns and Aspirations in the Cloud Ecosystem." Deliverable 7.1.2, December 2010b.

—. "Impact of OPTIMIS on Market Behaviour." Deliverable 7.1.3, July 2011a.

—. "Cloud Legal Guidelines." Deliverable 7.2.1.2, May 2011b.

Qiu F., Lin C. and Yin H. "EKM: An Efficient Key Management Scheme for Large-Scale Peer-to-Peer Media Streaming." Advances in Multimedia Information Processing PCM, 395–404, 2006.

Rafaeli S. and Hutchison D. "A Survey of Key Management for Secure Group Communication," *ACM Comput. Surv.* 35 (2003): 309–329.

Rodero-Merino L., Vaquero L.M., Gil V., Galán F., Fontán J., Montero R.S. and Llorente I.M. "From Infrastructure Delivery to Service Management in Clouds." *Future Generation Computer Systems* 26 (8) (2010): 1226–1240.

CHAPTER SIX

TCLOUDS: PRIVACY AND RESILIENCE FOR INTERNET-SCALE CRITICAL INFRASTRUCTURES

ALYSSON BESSANI, IMAD M. ABBADI, SVEN BUGIEL, EMANUELE CESENA, MINA DENG, MICHAEL GRÖNE, NINJA MARNAU, STEFAN NÜRNBERGER, MARCELO PASIN AND NORBERT SCHIRMER

Internet-scale information and communication infrastructures, also called infrastructure clouds, promise scalable virtualised computing, network, and storage resources over the Internet. They provide scalability and cost-efficiency but pose significant new privacy and resilience challenges since the pervasive clouds scenario may evolve into a single point of failure, threatening all interdependent information and communication infrastructures. These and other problems related with the security and resiliency of clouds are being dealt with by the EU-funded TClouds integrated project. This project proposes an extensible architecture integrating a set of technologies to build trusted cloud infrastructures as well as trusted services composed by untrusted commodity clouds resources. These technologies are being tested around two cloud-based critical applications: health and wellness monitoring and smart light management.

1. Introduction

Protecting critical infrastructures providing communications, energy, financial services or health-care presents increasing ICT (Information and Communication Technologies) challenges, as ICT itself has become vital to them. Internet-scale ICT infrastructures (infrastructure clouds) promise scalable virtualised computing, network, and storage resources over the Internet. They provide scalability and cost-efficiency but pose significant new privacy and resilience challenges besides the ones current critical infrastructures already have to deal with (Bessani et al. 2008; Nicol 2011). It is possible that clouds may evolve into a single point of failure, threaten all dependent ICT, and put the Future Internet at risk.

The TClouds project (http://www.tclouds-project.eu/) was created to address these concerns and increase the current understanding of the potential for commodity cloud-based services to host Internet-scale ICT infrastructures and how cloud-like infrastructures can be built to host these critical systems. In order to enable this vision, the TClouds project is built around a set of principles:

- **Organisation of independent (non-cooperating) providers:** One goal is to enable an ecosystem of independent providers. This will guarantee that the overall service does not depend on any individual provider.
- **Flexible trust models:** Trust models need to be flexible, i.e. users need to be able to determine what players and what mechanisms to trust. Another aspect is "user-centric security and privacy policies," i.e. those stakeholders using cloud-based services, be they individuals, SMEs or other organisations, can determine their individual security and privacy preferences.
- **Open interfaces:** In order to foster the creation of an adequate ecosystem for critical applications, openness of the architecture and interfaces is essential. This will enable interoperation between different clouds as well as seamless migration of workloads while maintaining the given security constraints.
- **Scalable security mechanisms:** Our security extensions must not break the underlying cloud principles. This means that deployed mechanisms must be scalable and transparent and must be resilient to expected failures of the underlying virtual infrastructure.

These principles are being applied on the TClouds security architecture, enabling the design and implementation of a set of middleware components that enable users to decide on their trust requirements, and encouraging user-centric deployment of appropriate security mechanisms trusted by individual users.

The TClouds architecture considers an organisation of trustworthy infrastructure clouds that build on complementary and mutually re-enforcing technical approaches:

(1) A Trustworthy Infrastructure Cloud enables individual providers to offer more resilient and privacy-aware infrastructure clouds

(2) Privacy and Resilience for Commodity Clouds enables end users to put a security layer on top of existing commodity infrastructure clouds to enforce their security objectives

(3) Federated Cloud-of-clouds Middleware privacy-protection and resilience beyond any individual cloud. This expands confidence from trusted (enterprise-internal) clouds to those less trusted (off-shored), or brings together a set of partially trusted providers into a trustworthy and adaptive federation.

An often overlooked problem that is also addressed in TClouds are the legal and business implications of the of privacy-enhanced cross-border infrastructure clouds.

All these aspects will be discussed in more detail in this chapter, organised in the following way. Section 2 describes security threats one has to face when using cloud-based infrastructures for hosting critical applications. Section 3 highlights some of the legal implications of using cloud infrastructures and platforms for hosting critical applications. Section 4 describes the application scenarios used to validate the TClouds architecture and components, which are described in Section 5. Section 6 discusses some basic issues and guiding principles for the TClouds security management framework, and Section 7 concludes the chapter.

2. Cloud Computing Security Threats

Cloud is a new term based on old technologies. Such technologies are well-established research areas and have their own security challenges inherited to cloud computing (Abbadi 2011; Abbadi and Lyle 2011). This section does not discuss security challenges related to individual cloud's structural component (e.g. virtual machine hypervisors, data center architectures, scalable storage architectures), but only security challenges that are a result of clouds properties.

Trustworthy clouds provenance. Logging, auditing and historical data are of tremendous importance in a cloud. This is especially the case as a cloud is expected to support Internet-scale critical applications. Logging, auditing and historical data have different usage, for example proactive service delivery (incidents monitoring and security monitoring), error investigation, billing, and forensic investigation. Almost all cloud resources generate this data in some way. At present, the only way that provenance is provided on a cloud is through linking together log and audit data, collected from multiple resources, to provide the complete history of an event or result. Most cloud infrastructures are composed of dynamically interlinked resources (for example hosted in different data centers). This means that building a logical sequence of events to investigate an incident for any one application requires data from many sources. These include the application itself, all logs for possible virtual resources that the application could have used, and logs of all physical resources that virtual resources could have used. Administrators must then combine this data correctly by identifying all time intervals when an application used a specific virtual resource, all possible time intervals when these virtual resources used physical resources and then all relevant log-files from all related resources. Collecting and combining data from these resources is not easy or practical considering the potential scale of cloud systems.

Secure infrastructure management. As discussed earlier, the cloud infrastructure is complex, heterogeneous and dynamic in nature. Considering such properties, it is inadequate to rely on human management of the infrastructure for the following reasons: (a) it requires more time to discover and resolve problems; (b) it requires more time to architect and deploy solutions; and (c.) it is error prone and does not provide a reliable way for measuring the level of trust in cloud operations, and is subject to insider threats by cloud employees. This raises the need to have trustworthy, self-managed middleware services that can automatically manage cloud infrastructural resources. Providing such management services raises many security challenges that we briefly discuss in Section 6.

Mitigating the risks of insiders. Insiders are cloud provider related users, who misuse their privileges intentionally or accidentally. Insiders have authorised credentials granted by a cloud provider to access the cloud infrastructure to perform their job functions. Insider threats could cause major impact on cloud operational services as well as on cloud user data. Some examples of possible insider threats are: (a) severe damage to cloud infrastructure when attacking automated self-managed services; (b) a cloud system administrator can connect to the hypervisor for a server, which hosts sensitive applications of a financial institution. From the hypervisor, the system administrator can access the memory space of the VM that runs the sensitive application (Rocha 2011), enabling them to access sensitive data; (c) a cloud system administrator using a system like

OpenStack (2011) can delete all VMs for an organisation and invalidate backups; and (d.) a cloud provider can copy VM images of one organisation onto a USB stick and sell it to a competitor.

User-centric security. Current cloud providers have full control over all hosted services in their infrastructure. For example, a cloud provider controls who can access VMs (e.g. internal cloud employees, contractors, etc) and where user data can be hosted (e.g. server type and location). The user has very limited control over the service deployment, has no control over the exact location of the provided services, and has no option but to trust the cloud provider to uphold the guarantees provided in the SLA. These issues raise several problems, which include the following: (a) regulation issues, especially when moving and processing data across territorial boundaries (see the next section); (b) user privacy and security concerns. For example, users need stringent assurance that their data is not being abused or leaked; and (c) how users can be assured that cloud provider properly isolate VMs that runs in the same physical platform (i.e. multi-tenant architecture [Ristenpart et al. 2009]) from each other. For example, two VMs for competing organisations might share the same physical platform. Such organisations need the assurance that others cannot intercept their applications and data.

Loss of availability. There are many reports of serious outages and problems that have made cloud-hosted applications unavailable for periods ranging from hours to days (see Raphael 2011 for a survey of the 10 worst cloud outages). This is a fundamental problem that cloud-users have to face when their infrastructure is outsourced: the dependability (in particular, availability) of the cloud-hosted service is not only dependent of the cloud provider infrastructure, but from the Internet itself. This means that, despite many providers claiming to provide services with five lines of availability, in practice the lower availability of the Internet makes the perceived availability of cloud-hosted services much smaller.

3. Legal Implications of Cloud Computing

One of the first aspects that the TClouds project is looking at are the legal challenges associated with the flexibility of cloud computing. The market-leading cloud providers rely on infrastructures that are usually spread across different countries. The dynamic trans-border transfer and storage of data pose several legal problems especially regarding data privacy.

The multinational infrastructure raises the question of what jurisdiction or law governing the handling of personal identifiable information (PII) is applicable. The comprehensive data protection legislation of EU member states is applicable whenever the provider (or a unit thereof) is located or uses equipment for data processing in a member state. Equipment in this context means any technical facility or personnel that the provider operates within the EU. Therefore, any European server farm or any European availability zone results in applicability.

The EU data protection framework restricts the transfer of PII to outside the EU. It demands adequacy of data protection standards of the recipient to guarantee sufficient safeguards of the individual's right to information self-determination. Only a few countries meet these high expectations of adequate privacy provisions. 75% of states worldwide do not have comprehensive data protection laws, including the USA.

Data transfer to these so-called third states is only allowed under very strict contractual conditions. The European Commission established a self-certification framework for US companies, the US-EU Safe Harbor Framework. This allows US companies to self-certify adherence to privacy principles that should provide adequate data protection standards within this company. However, this framework has been strongly criticised as insufficient from the European point of view (Commission 2004).

Further contractual enablers for third state data transfer are the EU Standard Contractual Clauses, as well as Binding Corporate Rules. However, experience with these contractual solutions in regard to cloud providers is short.

TClouds analyses these conditions to provide guidelines on additional measures to enable lawful cloud computing. This could be either comprehensive encryption, the reliance on cloud providers solely located in the EU or the adjustment of contractual solutions to the needs of cloud customers. It is worth noting that the legal requirements of some of mentioned solutions are supposed to address interesting research problems. For instance, genetics-related data requires long-term encryption to ensure owner privacy even after a person dies, to protect the privacy of their descendants.

4. Application Scenarios

The TClouds project guides its development and evaluates its results by means of real-life use cases. These use cases provide a set of critical infrastructures that will demonstrate the privacy-enhanced resilience provided by our infrastructure. In this section, we present the two main application scenarios considered in the project.

Health monitoring. The TClouds medical use case is a home healthcare application that is particularly designed to help depressed patients during their therapies. Patient information such as daily mood diaries, physical activities, sleep and light monitoring information that is proven to have high impact on the cure of the disease, are continually remotely monitored, uploaded to the cloud, and shared among different healthcare service providers. As depicted in fig. 6.1, the actors involved with the home monitoring scenario are the patient, their family, and the psychiatrist. They interact with the proposed system and access the allocated services via interfaces provided in web portals. Two service providers are involved: the hospital service provider that provides EHR (Electronic Health Record) and PHR (Patient Health Record) based services, and the health and wellness service provider that provides physical activity and sleep monitoring services through dedicated home monitoring devices.

Fig. 6.1. Home healthcare application scenario and actors

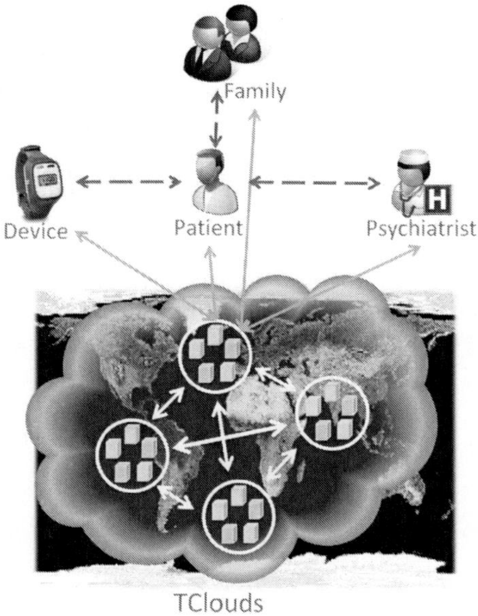

A noteworthy aspect of this application scenario is that the service providers here are not from the IT business, so potentially they will have great benefits in outsourcing the infrastructure of the service to some cloud providers. However, there are many critical non-functional requirements for this kind of application that unfortunately no cloud provider can currently satisfy. First, the monitoring service must always be available, something that no cloud provider can ensure (see Section 5.2.2). Second, the stored data must be confidential and the patient privacy should be preserved, since we are dealing here with PII (Personal Identifiable Information) subject to the European data protection legislation (see Section 3). This conflicts directly with the use of commodity clouds since most of them, store data outside of Europe.

Smart public light management. Smart grids add intelligence to the energy distribution infrastructure. They are built to reduce outages and faults, improve responsiveness, increase efficiency and reduce management costs. A smart grid uses sensor meters, digital controls and analytic tools to automate and monitor energy flows. It enables energy and information flow between generators, consumers and any intermediary entity. New sustainable distributed power sources such as wind and solar generation or mobile, plug-in power drains such as electrical vehicles can only be widely integrated using smart grids.

TClouds proposes a use case for smart lighting as part of a smart grid, providing "public lighting management" with switching commands (on-off), real time status, energy consumption and schedules update. Accuracy and timeliness are crucial here, because decisions are taken in real time. No information and information at the wrong time represent inaccurate analysis that will turn into bad decisions. To prevent it, information must be suitably protected, especially in the interconnected clouds.

Fig. 6.2. Smart Lighting Architecture

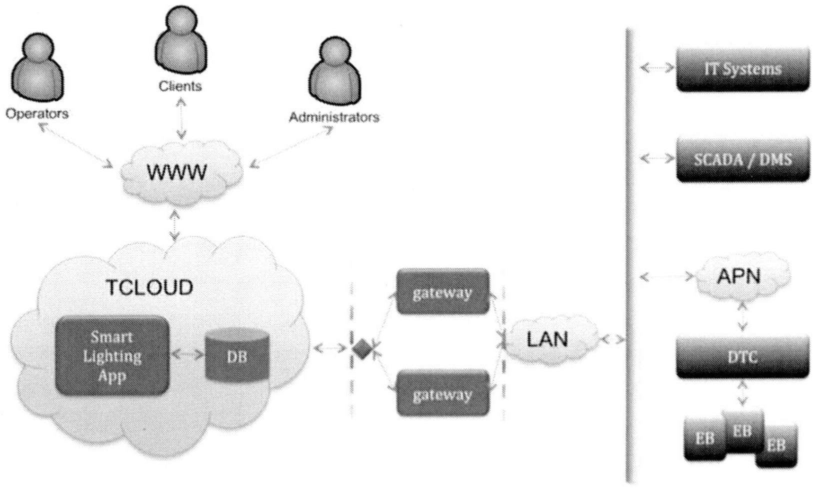

The TClouds' smart lighting system main components are displayed in fig. 6.2 above. In the figure, IT Systems are the existing systems such as metering and sales. SCADA/DMS (Supervisory Control and Data Acquisition/Distribution Management System) are control systems for the supervision, optimisation and management of power distribution. DTCs (distribution transformer controllers) are equipment installed at switching stations, including modules for measuring and actuation. EBs (energy boxes) are devices installed at consumers and producers, including modules such as for measuring and actuation.

Smart grids have the same information security needs of current electrical and telecommunications sectors. They have systems to deal with millions of customers, highly critical assets for the functioning of society, storing sensible customer information, and potentially generating detailed information about their habits. In particular, a smart lighting system does not have critical data related with customers, making data confidentiality issues less critical than data integrity, availability or authenticity.

5. TClouds Integrated Architecture

In this section, we present the TClouds integrated architecture that encompasses all of the principles discussed in the introduction, presenting alternative deployments for this architecture. In short, TClouds can be described as a resilient cloud-of-clouds infrastructure providing automated computing resilience against attacks and accidents, complementing or in addition to commodity clouds. This enhanced functionality will be achieved through specialised middleware standing between low-level, basic multi-cloud untrusted services, and the applications requiring security and dependability.

The TClouds architecture supports implementations of TClouds functionality preserving the use of legacy commodity clouds IaaS, either by resorting to client-side software, or to server-side software. Moreover, the architecture allows for more ambitious steps, those considering that commodity cloud providers will eventually adhere to a model such as TClouds, directly providing resilient cloud computing.

The architecture, shown in fig. 6.3 below, provides applications with a wealth of interfaces to produce incremental resilience solutions with single or multiple clouds: TClouds Trusted Platform services (T-PaaS*)* on top of the middleware layer, TClouds Trusted Infrastructure services (T-IaaS) from within the middleware layer, IaaS from available commodity untrusted clouds. The main building blocks of the architecture are briefly described in the following.

Basic multi-cloud untrusted services represent the available standard functionality, at IaaS level, offered by commodity market players. Available services may evolve with the evolution of these systems, but are normally confined to processing power, storage, networking, and several input/outputs.

Trusted infrastructure services represent trusted-trustworthy versions of IaaS services, namely storage and processing power. The idea is to offer resilient file systems, and low-level virtual machines, by combinations of fault, intrusion prevention, tolerance mechanisms and protocols which build a resilience layer on top of the corresponding untrusted storage and processing systems. An example of this kind of solution are the trusted virtual domains, presented later on this section.

Fig. 6.3. TClouds building blocks

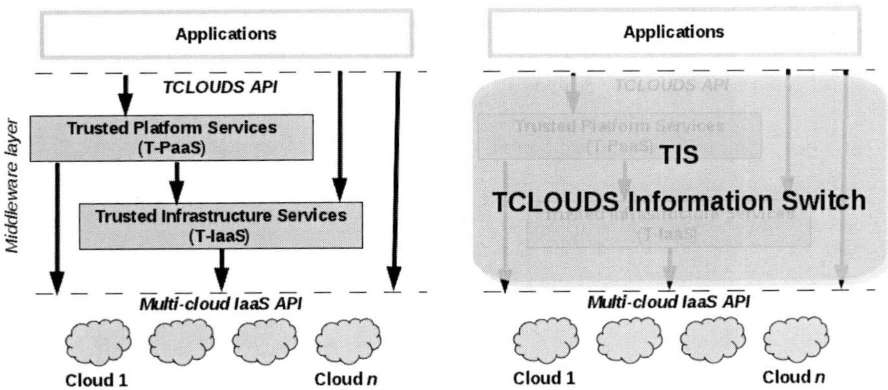

Finally, trusted platform services represent trusted-trustworthy services at a higher level of abstraction, provided through extensions of the resilience layer implemented by the TClouds middleware, built on top of either or both the commodity IaaS clouds and the T-IaaS. These services normally deploy semantics useful to build complex reliable and distributed applications, as those described in Section 5.2.2.

In order to adapt this architecture to different deployment alternatives that we shall discuss later in this section we introduce the notions of TClouds Information Switch and TClouds Information Agent.

TClouds Information Switch (TIS) is a conceptual block that runs the middleware protocols and mechanisms implementing the resilience components already mentioned, and see also fig. 6.3 above. Each TIS instantiation encapsulates the services in use by that configuration, which are all or part of the services defined in the TClouds architecture. TIS implementation depends on the particular incarnation, ranging from dedicated machine to fault and intrusion tolerant appliance box containing several TIS replicas implemented as virtual machines. The TIS can be built with incremental levels of resilience, depending on its criticality. Trustworthy TIS-TIS interconnection through TClouds communication services secures information flows in the architecture.

TClouds Information Agent (TIA) can be seen as a particular implementation of a TIS, as a software appliance residing with end clients. Like the TIS, it runs different sets of functions, depending on specific protocols being used on the client side. TIAs require no additional hardware as a general rule. However, running in the client space, they are subject to a great level of threat. This can be mitigated by configurations where the TIA logic is aware of the existence of minimal additional hardware (for example trusted components) to improve its resilience. On the other hand, the TIA option requires client modifications to achieve the desired TClouds functionality. Whenever needed, trustworthy TIA-TIS interconnection through TClouds communication services secures information flows in the architecture.

In fig. 6.4 we present a rich infrastructure prefiguring a true ecosystem capable of offering the best possible tradeoffs to clients and providers of resilient cloud services, either end-clients or mediators.

Fig. 6.4. TClouds architectural options

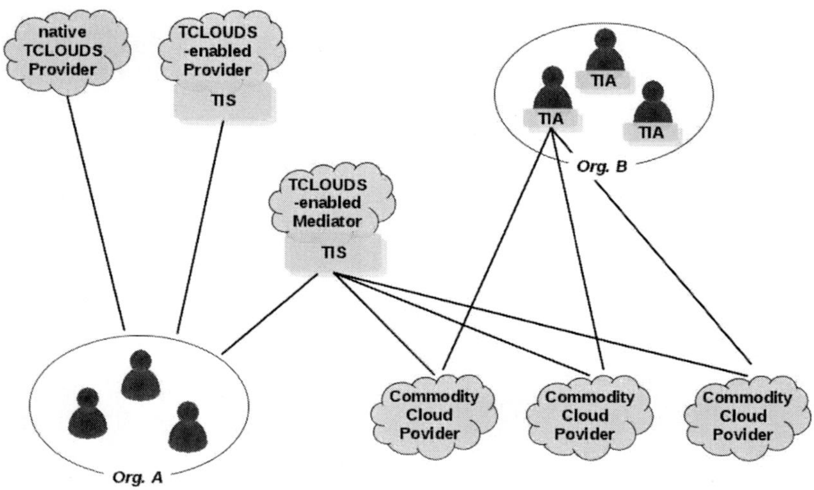

From left to right, we can see end-clients from an organisation (Org. A), that access "native TClouds" or "TClouds-enabled" cloud providers. The same—unmodified—clients can access resilient cloud services implemented on top of commodity clouds through a "TClouds-enabled mediator." Finally, another way to achieve resiliency from commodity clouds is via TClouds-enabled client-resident software, as shown by users from the other organisation (Org. B) in fig. 6.4 above.

In the remainder of this section, we present specific contributions of the project towards the development of the architecture. We start by describing how T-IaaS can be built using trusted components and virtual machines technology.

5.1 Building Trusted Clouds

When designing a trusted cloud infrastructure from scratch we want to meet the core requirements below, for a secure and trustworthy cloud.

(1) **Separation of duties:** we distinguish between the duties of the cloud administrator (provider) and the organisation administrator (customer). The cloud administrator works for the provider and is responsible for setting up and managing the resources of the cloud provider (servers, storage, etc.). In contrast, the organisation manager is responsible for managing the virtual infrastructure and security policies of the customer.

(2) **Isolation:** in a multi-tenant cloud scenario a strict and strong isolation between the virtual infrastructures of the different customers is of utmost importance. Moreover, isolation from the cloud administrators should be achieved. Confidential data of the customers should not be accessible to cloud administrators, other customers or any other entities except those authorised by the customer.

(3) **Verifiability of the platform:** as the cloud resources run off-premise there should be technical means to measure the integrity of the remote platform. Thereby a technically enforceable level of trust can be established which goes beyond audits, certification or service level agreements.

(4) **Integration into organisation infrastructure of the customer:** the cloud resources are accessed via various user clients ranging from mobile devices, desktop computers and on-premise servers of the cloud customer. We need to establish a seamless integration of the cloud resources into the organisations infrastructure and

maintain a consistent security level among all resources. Only then will we have reliable end-to-end security. We propose a central management component that is under the control of the organisation.

The basic functionality of a server component within the cloud infrastructure is to provide virtualisation via a hypervisor. The deployment of virtual machines (VMs) to the server is managed by a cloud management layer, as show in fig. 6.5 below.

Fig. 6.5. Today's cloud infrastructure

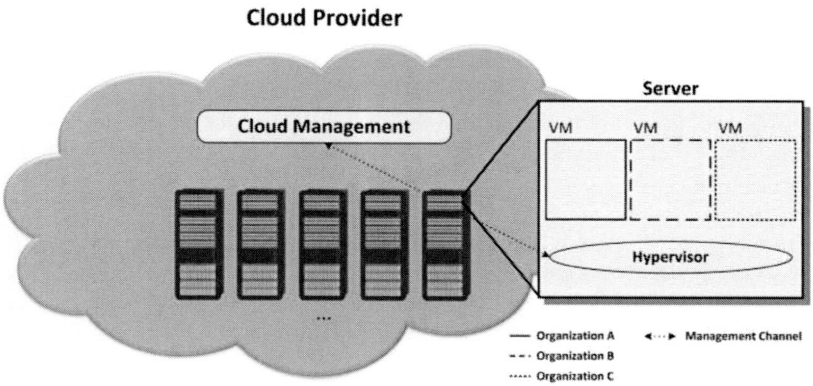

As a server in a public cloud can run VMs of different customers (which may be competitors), the isolation of the VMs is of utmost importance. Beyond the memory isolation a hypervisor offers, we demand customer centred encryption of the data as it is stored on virtual hard drives or databases. However, during computation on the user data it is processed in plain text within the memory of the VM. Hence, we need technical means to limit the power of the cloud administrator to use the hypervisor to look inside the memory of the VMs. To deal with this insider threat we propose the use of a strict interface of the hypervisor that only allows an administrator to start, stop and migrate VMs, but not to dump the memory. The hypervisor should be supervised by a security kernel (Sirrix AG 2008) that provides a clear management interface (for remote management) and enforces isolation and encryption on the server. The security kernel moderates all management aspects and there is no need for a so called "root" account with full access to resources for cloud administrators. In addition, the security kernel can provide mechanisms for logging. In the following, we refer to the combination of the hypervisor and the security kernel as secure hypervisor.

To ensure the integrity of the server hardware and the security kernel we propose to employ Trusted Computing technology by the Trusted Computing Group (2011a). A Trusted Platform Module (TPM) (Trusted Computing Group 2011b) is built into the server and is the hardware anchor for the root of trust. During the boot process of the server the TPM measures the state of the hardware and the operating system (hypervisor and security kernel) and compares these to the valid configurations it expects. Only if the measurements match the expectations of the TPM the server will be started. Moreover, the TPM allows a remote party (the cloud customer) to verify the integrity of the server via remote attestation. This is the basis for verifiability of the cloud platform's integrity by the cloud customers. Moreover, we employ the TPM to build up a trusted channel (encrypted, mutually authenticated and mutual integrity check via remote attestation) between the trusted cloud management component and the secure hypervisor, as show in fig. 6.6 below.

Fig. 6.6. Trusted cloud infrastructure with TPM, secure hypervisor and trusted (remote) cloud management

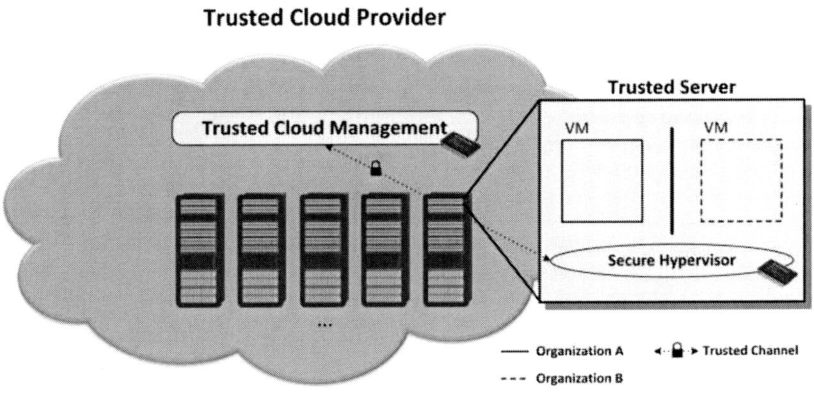

Up to now, the secure hypervisor isolates the VMs of different customers. To properly deal with realistic security requirements, we must go one step further and introduce a finer grained compartmentalisation representing the different security domains within the organisation of the customer. Typical examples for such security domains are the departments of an organisation such as human resources, sales, accounting, management and development. Those departments deal with different aspects and workflows of an organisation, are typically operated by different employees (although there might be some overlap), and have different security requirements.

Hence, these domains should also be isolated and the information flows should be controlled by the security kernel according to the organisation's security policy. This concept is known as *Trusted Virtual Domains* (TVDs) (Griffin et al. 2005). It allows the deployment of virtual domains on shared resources (including computing, storage and networking). Computing resources are typically encapsulated in a VM and a running instance (called compartment) is assigned to a single TVD. All data within a TVD is bound to that TVD by cryptographic means. Hence, data at rest is protected to stay within the TVD. Compartments of the same TVD can communicate with each other while being isolated from other TVDs. Remote communication between compartments over untrusted networking infrastructure is properly secured by the security kernel. This allows the secure integration of remote branches of an organisation into a single virtual infrastructure. Finally, every information flow between TVDs is also controlled by the security kernel and is only granted if it complies with the security policy.

This scenario with TVDs fits perfectly to the trusted cloud infrastructure just described. As the trusted servers already have TPM support and a security kernel in place they can be extended with the TVD concepts. To ensure a homogeneous security level between on-premise resources of the customers and cloud resources we opt for a central management component for organisation's TVDs, the Trusted Organisation Manager. This management component is controlled by the organisation (or a trusted third party) and manages all of its appliances.

The security policy defining the TVDs and information flows allowed is managed by the central component and communicated to the security kernels of the appliances via a trusted management channel. Moreover, the management component of the organisation also communicates the security policy to the trusted cloud management component (also via a trusted management channel). It is the duty of the trusted cloud management component to distribute this policy to the trusted servers. With this architecture, the security policy and the on-premise appliances are still under the control of the organisation's management component, while the enforcement of the security policy is delegated to the cloud components where necessary. In fig. 6.7 we sketch such a scenario where VMs of different TVDs of an organisation are deployed in the cloud side by side with VMs of other customers. They are securely connected to on-premise VMs in the same TVDs. Similar to a trusted server the trusted desktop appliance ensures isolation of TVDs by a secure hypervisor and is managed by the same trusted organisation manager.

Fig. 6.7. Trusted virtual domains spanning internal infrastructure and a public trusted cloud

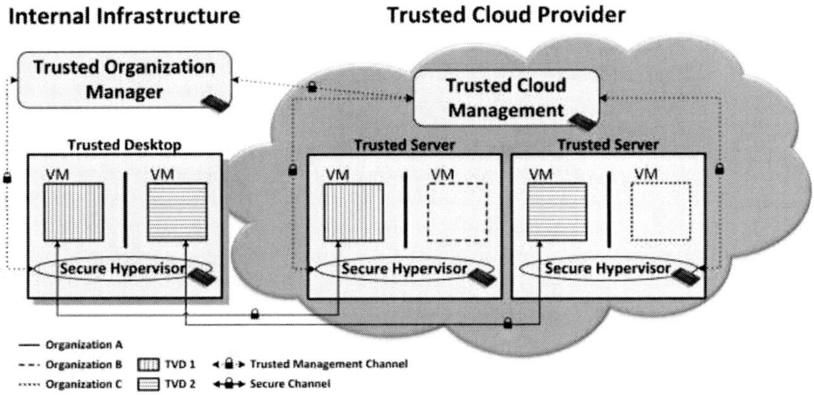

5.2 Trusted Cloud Services from Commodity Clouds

Today there are many public clouds available on the Internet providing elastic services ranging from arbitrary virtual machine execution (e.g. Amazon EC2, Rackspace CloudServers, IBM Smart Cloud) through object storage (e.g. Amazon S3, Rackspace Files, Google Docs), passing through web services platforms (e.g. Google AppEngine, Amazon Elastic Beanstalk), databases (e.g. Amazon RDS and Windows Azure SQL Server) specific services such as message queuing (e.g. Amazon Queue Service) and email (e.g. Gmail).

There are three big advantages of using these clouds when compared with more restricted private, albeit trusted, clouds. First, they offer a model where you pay for what you use, so you do not need big investments to start benefiting from the clouds. Second, they provide an almost unlimited amount of resources, allowing applications to scale without bound (in terms of resources). Third, they exploit the economy of scale (many clients and many resources) to provide resources and services in a very cost-effective way.

However, as discussed in Section 2, the public clouds suffer from serious drawbacks when considering their use for critical applications. In fact, the existing technology demands a client to completely trust the cloud provider when deploying its application. This trust is hardly justified since problems like vendor lock-in, outages and breakdowns (Raphael 2011) and insider attacks (Rocha 2011) are real threats to the dependability requirements of critical applications.

In the following, we discuss some approaches being taken in the TClouds project to implement and/or understand the limitations of implementing trusted cloud computing services using untrusted commodity clouds. However, before presenting these solutions, we discuss to what extent the use of cryptography is effective in augmenting the security of (commodity) cloud computing.

5.1.1 The Role of Cryptography in Untrusted Clouds

As cloud computing, especially IaaS, offers the ability to run commodity software in the cloud, it seems natural to apply already existing solutions for security as well. Otherwise, when handling sensitive data in the cloud, the client has to trust the cloud provider. Classical cryptography, from the pre-cloud-computing age, can be used to accomplish confidentiality. However, as already discussed, in order to process data, it has to be decrypted at some point. Plaintext data processed inside the VM instances is then temporarily stored in the VM's memory. One possibility to ensure confidentiality nevertheless would be to perform all operations under encryption (for example using Fully Homomorphic Encryption), which however, is currently not yet efficient in practice. On the other hand, when using classical cryptography, we have to assume that nobody but the customer can access the VM's memory. That implies that cloud administrators have only remote access with limited privileges and cannot inspect memory. (The TVD solution discussed in previous section solves this problem, but it requires the availability of trusted hardware—TPM—which, to the best of our knowledge, is currently unavailable in all commodity clouds.) These implications are depicted in fig. 6.8 below, and discussed in the following.

Fig. 6.8. Trust models and their implications

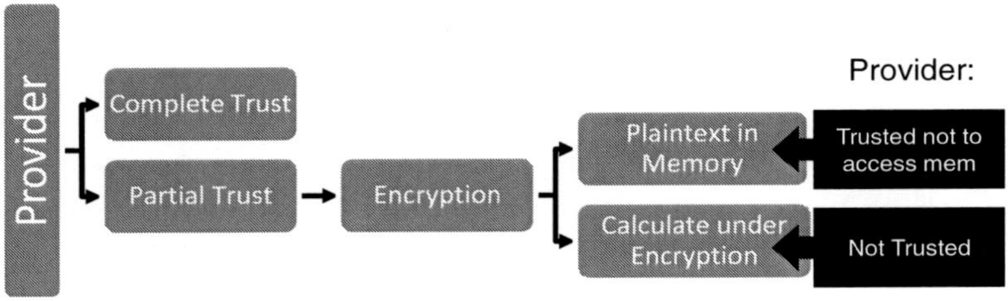

Calculate under memory (the paranoid approach). A possible solution for operating under encryption in the cloud is TwinClouds (Bugiel et al 2011a). TwinClouds proposes a model that uses Garbled Circuits—binary non-recyclable circuits—that can calculate on encrypted data and output encrypted data again without the need to ever decrypt in an intermediate step. However, garbled circuits can only be used once and their creation is costly. The TwinClouds model therefore accumulates secure but slow computations over time and provides the possibility to query them in parallel on demand by leveraging the benefits of cloud computing. In this approach, the customer communicates with a resource-constrained "trusted cloud" (either a private cloud or built from multiple secure hardware modules) that encrypts algorithms and data to be stored and later on queried in the powerful but untrusted commodity public cloud.

Plaintext in memory (the "keys in the cloud" approach). When not using the aforementioned operation on encrypted data, another possibility is to encrypt sensitive data and trust the provider not to read memory contents while operating on the data (see figure 6.9 below). If we trust the provider not to eavesdrop on memory, the question "why not store unencrypted data in the memory in the first place" remains. The answer is quite simple: cloud computing is designed to run multiple instances that might access a common data source, even outside the cloud, which needs to remain confidential. The necessary keys for the needed cryptography have to be stored inside the VM somehow. Controlling the use of the same key in different instances or managing revocations is hard or even impossible. Further, provisioning this key is only possible by placing it in the VM image or by sending it to the running VM instance over a trusted channel. To establish a trusted channel in turn, a secret key already in place in the VM image is needed for authentication. Putting the key in a VM image is however as secure as writing your PIN on your cash card. The external storage, where a dormant VM image is saved, is easier to access than a running VM instance's memory and faces the risk of lost control once the physical hard drives have crashed or have been replaced. Moreover, keys stored in a VM impede the concept of VM image sharing and the needed cryptography needs to be implemented and managed by the cloud's customer. Cryptography in combination with VM sharing is explained in the next subsection. The different alternatives of storing keys and their implications are depicted in fig. 6.9 below. As can be seen in this figure, all solutions require some trust on the provider.

Fig. 6.9. Possible positions of keys

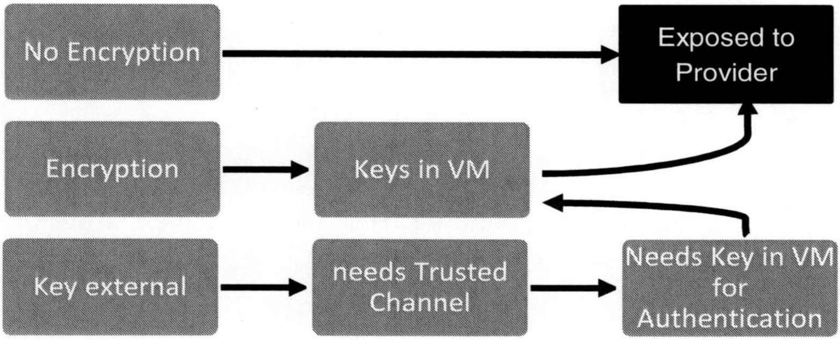

Careless usage by novices. The rise of cloud computing also brought complementary products like Appliance Store with it (for example the Cloud Market http://www.thecloudmarket.com). Comparable to App Stores for smartphones, the appliance stores for cloud computing enable customers to build complete virtual machines (appliances) and share them with literally thousands of other customers. As easy as this solution might seem at first sight, this procedure carries a significant privacy risk with it, as it is comparable to freezing the current state of your computer and sharing it with thousands of unknown people. An obvious assumption was found to be true by TClouds researchers (Bugiel et al. 2011b): almost a third of all tested shared Amazon Machine Images (AMIs) contained private data. This does not only include credentials or source code, but even worse: cryptographic keys that enable the opening of (believed) protected data outside of an AMI. They also found private keys and certificates for authentication that would allow an adversary to impersonate the identity of a certified web server. Furthermore, due to cloning of the process of sharing VMs, identification keys like SSH host keys are cloned, too. This does not only enable the identification of thousands of potentially vulnerable hosts but an inadvertently cloned user accounts is practically a back door to all cloned VMs. These vulnerabilities are rooted in the careless and error-prone manner in which Amazon's customers handle and deploy AMIs.

In conclusion, the solutions and threats discussed in this section show that storing keys and other security critical data in the cloud by default require complete trust in the provider. In the next section we discuss a solution for a TIA (TClouds Information Agent) that allow the use of a set of untrusted clouds to store private information.

5.1.2 Cloud-of-Clouds Services

One of the research directions of TClouds is to show how to combine several clouds to build trusted cloud services using resources from a diverse set of commodity public clouds, employing a paradigm known as "distributed trust" (Schneider and Zhou 2005) to implement a "cloud-of-clouds" able to tolerate faults and intrusions on up to a certain number of providers. The rationale of the approach is that by exploiting replication in diverse and non-correlated infrastructures such as those provided by different commodity cloud providers, one can avoid both accidental and malicious failures.

There are several challenges that need to be addressed to make this approach work. More precisely, it requires replication mechanisms that (a) can tolerate a large spectrum of faults and security threats, (b) can cope with Internet communication unpredictability to effective the smooth running of multi-cloud systems, (c) are compatible with most public clouds available without changing their infrastructures or require their cooperation, and (d) minimise the replication resource overhead to make the solution economically viable for practical applications. These challenges cannot be completely addressed by state of the art replication protocols and thus new techniques must be devised.

An example of a service being developed at TClouds is the DepSky cloud-of-clouds object storage system (Bessani et al 2010), which uses object storage services from diverse cloud providers (e.g. Amazon S3, Rackspace Files) to build virtual trusted object storage. Fig. 6.10 below illustrates the basic idea behind DepSky.

Fig. 6.10. DepSky cloud-of-clouds architecture

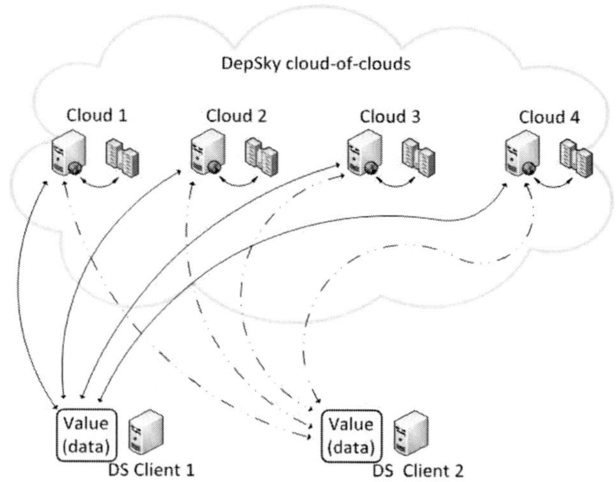

The core of the solution is a set of read/write protocols based on the use of Byzantine quorum replication (Malkhi and Reiter 1998) requiring *3f+1* clouds to tolerate up to *f* cloud faults. These protocol address the requirements mentioned in the following way:

(1) The system tolerates arbitrary (a.k.a. Byzantine) faults in order to cope with all possible behaviours of a fraction of providers.

(2) The replication protocols operate on an unreliable network in which messages can be lost and delayed and do not require participation of the full set of employed clouds, but only of a subset of them (a quorum), on any step of the protocol execution.

(3) The protocols are completely client-based, in the sense that no specific cloud on the clouds is required. DepSky assume the clouds provide storage service with standard operations for managing objects and containers (put, get, list, etc.). Moreover, the set of storage clouds do not interact among themselves, but only with the clients.

(4) The storage protocol provided by DepSky employs RAID techniques such as "erasure codes" (Rabin 1989) to store only a fraction of the data on each cloud, making the solution cheaper than what would be expected for a full replication scenario (cost less than *N* times as much as single cloud storage, *N* being the number of clouds in which the data is replicated).

DepSky employs a "cryptographic secret sharing scheme" (Shamir 1979) to ensure that no single cloud can obtain any information on the stored data. The idea is to make the information stored readable only if the object is read from more than *f* clouds, solving the problem of confidentiality of stored data.

The big difference between DepSky and some services that provide secure backup or data sharing (for example Wuala http://www.wuala.com/) is that in the latter, stored data owners need to completely trust the "secure" service provider, while DepSky requires no trust in individual parties.

A prototype of DepSky was deployed in four public clouds, namely Amazon S3, Rackspace Files, Windows Azure Blob Service and Nirvanix CDN, forming a cloud-of-clouds, accessed periodically by several clouds around the world for a month. The main results of the experiment were the following. In terms of monetary costs, the use of erasure codes makes DepSky storage cost from 23% to 50% more than single cloud storage (on average). In terms of performance, the DepSky read protocol performance have smaller latency than any single cloud (data is fetched from the faster cloud at the moment) while the write protocol latency is close to the worst cloud (data needs to be written on a quorum of clouds). This is may not be a problem since most storage systems tend to be exposed to read-dominated workloads. However, we are yet to see if this is the case for TClouds application scenarios.

Currently, in TClouds, we are working on the evolution of the DepSky prototype to make it usable in project applications (especially in the healthcare scenario) as well as in the implementation of a cloud-of-clouds file system, a DropBox-like service, and a secure backup system, among other things. Moreover, we are still actively working on alternative protocols for cloud-of-clouds storage that explore regions of design space different from the ones explored by DepSky.

The DepSky system deals with the problem of data replication using a cloud-of-clouds. A more complex problem we are currently addressing is the how to use the computing resources available on public clouds to build dependable cloud-of-cloud computing services. Current work in this field revolves around the development of a T-PaaS built using virtual machines deployed on a diverse set of cloud providers. The technical challenges of providing a cloud-of-clouds computing platform are substantially harder than for the storage case. The main reason comes from the fact that the required protocols for general service replication (called state machine replication) usually do not work very well on WANs like the Internet, so new protocols such as EBAWA are under development (Veronese et al. 2010).

Besides the aforementioned services for storage and computing, there is work being done on building a secure log service replicated among a cloud-of-clouds. One of the initial ideas is to use a service like DepSky to ensure that the log registers cannot be forged.

6. Managing the Security

Managing cloud security is an important but a complex subject to tackle, considering its complexity and dynamism. Security management, for example, covers managing protection keys of users' data and infrastructural management data, secure policy management and enforcement, and secure operational management. Secure operational management is the foundation for managing the security of clouds.

In TClouds, we are mainly interested in the following two elements: (a) supporting self-managed services to securely manage virtual layer resources at physical layer resources (i.e. VM creation and migration) based on both user properties and infrastructure properties; and (b) enabling users to assess the trustworthiness of such management services. Point (a) requires establishing trustworthy and secure infrastructure across clouds' distributed entities, while point (b) involves establishing a chain of trust between cloud users and the resources at clouds infrastructure. Cloud services should provide transparent infrastructure management to cloud users. Therefore, a chain of trust needs to be established between users and management agents (for example OpenStack), and another chain of trust between these agents and cloud's infrastructural resources.

The provision of secure and trustworthy infrastructure for the operation of self-managed services requires the following mechanisms: (a) VMM trustworthiness attestation to ensure that it would enforce user properties; (b) communication of user properties across cloud related components, ensuring these properties are not tampered with whilst being transferred, executed or stored; (b) provision of secure information sharing across cloud components in the same layer and across multiple layers; (c) mitigation of insider threats; (d) Standardisation—most technologies used in cloud are not new; however, the heterogeneous nature of the cloud requires the reconsideration of many issues, as in the case of standardisation. For example, different software and hardware providers need to provide standard interfaces enabling cross communication between cloud components; finally, (e) Interoperability—this requirement is not only to avoid vendor lock-in, but also to enable collaborative efforts. For example, hypervisor and VMM interoperability enables VMs from different suppliers to work on hypervisors from different manufacturers. Notice that most of these requirements are provided by the trusted virtual domain implementation discussed in Section 5.1.

These concerns are mostly related with the management of trusted clouds. When one wants to deploy security mechanisms on commodity clouds, such as in the case of TwinClouds (Section 5.2.1) and DepSky (Section 5.2.2), the management capabilities are reduced only to what the cloud operator exposes to the cloud user. This severely constrains the management options of the service user, and thus requires the specific commodity cloud-based service to deal with this limitation, possibly under its TIA. One of the most important challenges yet to be addressed in TClouds is how to manage the security heterogeneous untrusted resources from commodity clouds.

7. Conclusions

In this chapter we have presented some of main concepts being explored in the TClouds integrated three-years project. At the time of writing, we are finishing the first year of the project and main focus is on the development of a single conceptual framework that integrates the many facets of the cloud security problem.

The work presented in this chapter does not represent everything being developed at the project, but only the more mature parts. For more information about TClouds, visit the project webpage http://www.tclouds-project.eu/.

Acknowledgments

We would like to thank the reviewers (specially the first) for the feedback and encouragement that helped not only this text but on refining some questions we need to answer during the next two years of the project. We also thank all TClouds project partners by the many fruitful discussions about the work presented here. Finally, we thank the second WoSS organisers and participants for their invitations and the many interesting discussions in Timisoara. This work is supported by European commission through project TCLOUD (FP7/2007-2013, ICT257243).

References

Abbadi Imad, M. "Toward Trustworthy Clouds' Internet Scale Critical Infrastructure." In *7th Information Security Practice and Experience Conference (ISPEC '11)*, volume 6672 of LNCS, pages 73-84. Springer-Verlag, Berlin, 2011.

Abbadi Imad, M. and Lyle, John. "Challenges for provenance in cloud computing". In *3rd USENIX Workshop on the Theory and Practice of Provenance (TaPP'11)*. USENIX Association, 2011.

Bessani, A. N., Sousa, P., Correia, M., Neves, N. F. and Veríssimo, P. "The Crutial Way of Critical Infrastructure Protection." *IEEE Security & Privacy*, IEEE Computer Society, 2008.

Bessani, Alysson, Correia, Miguel, Quaresma, Bruno, André, Fernando, and Sousa, Paulo. "DepSky: Dependable and Secure Storage in a Cloud-of-Clouds." In *6th ACM SIGOPS/EuroSys European Systems Conference (EuroSys'11)*, 31–45, 2011.

Bugiel, Sven, Sadeghi, Ahmad-Reza, Schneider, Thomas, Nürnberger, Stefan. "Twin Clouds: Secure Cloud Computing with Low Latency." In *Communications and Multimedia Security Conference (CMS'11)*. New York: Springer, 2011a.

Bugiel, Sven, Pöppelmann, Thomas, Nürnberger, Stefan, Sadeghi, Ahmad-Reza and Schneider, Thomas. "AmazonIA: When Elasticity Snaps Back." In *18th ACM Conference on Computer and Communications Security (CCS'11)*, 2011b.

Commission Of The European Communities. European Commission Staff Working Document: "The implementation of Commission Decision 520/2000/EC on the adequate protection of Personal data provided by the Safe Harbour privacy Principles and related Frequently Asked Questions issued by the US Department of Commerce." SEC (2004) 1323. Brussels, October 20, 2004. http://ec.europa.eu/justice/policies/privacy/docs/adequacy/sec-2004-1323_en.pdf (accessed September 15, 2011)

Griffin, J. L., Jaeger, T., Perez, R., Sailer, R., VanDoorn, L. and Caceres, R. "Trusted Virtual Domains: Toward secure distributed services". In *1st Workshop on Hot Topics in System Dependability (HotDep'05)*, 2005.

Malkhi, D. and Reiter, M. "Byzantine Quorum Systems." *Distributed Computing* 11 (4) (1998): 203–213.

Nicol, D. "Hacking the Lights Out: The Computer Virus Threat to Electrical Grid" *Scientific American*, 2011.

http://www.scientificamerican. com/article.cfm?id=hacking-the-lights-out (accessed September 15, 2011)

Rabin, Michael. "Efficient Dispersal of Information for Security, Load Balancing and Fault Tolerance." *Journal of the ACM* 36 (2) (1989): 335–348.

Raphael, J.R. "The 10 worst cloud outages (and what we can learn from them)". *Inforworld*, 2011. http://www.infoworld.com/d/cloud-computing/the-10-worst-cloud-outages-and-what-we-can-learn-them-902 (accessed September 15, 2011)

Ristenpart, Thomas, Tromer, Eran, Shacham, Hovav and Savage, Stefan. "Hey, You, Get Off of My Cloud: Exploring Information Leakage in Third-Party Compute Clouds". In *16th ACM conference on Computer and Communications Security (CCS '09)*, 199–212, 2009.

Rocha, Francisco, Correia, Miguel and Abreu, Salvador. "Confidentiality and Privacy in the Final Frontier: Inside the Clouds." *IEEE Computer*, IEEE Computer Society, 2011.

OpenStack. "OpenStack Compute—Administration Manual." 2011. http://docs.openstack.org (accessed September 15, 2011)

Schneider, Fred B. and Zhou, Lidong. "Implementing Tustworthy Services Using Replicated State Machines." *IEEE Security & Privacy* 3 (5) (2005): 34–43.

Shamir, Adi. "How to share a secret." *Communications of the ACM* 22 (11) (1979): 612–613.

Sirrix, AG. "High-Assurance Security Kernel Protection Profile (EAL5)." According to the Common Criteria v3.1 R2, 2007, certified by German Federal Office for Information Security (BSI), 2008.

Trusted Computing Group. "Trusted Computing Group webpage." 2011a. http://www.trustedcomputinggroup.org (accessed September 15, 2011)

Trusted Computing Group. "TPM Main Specification, Version 1.2, Revision 116." 2011b.

Veronese, Giuliana, Correia, Miguel, Bessani, Alysson and Lung, Lau. "EBAWA: Efficient Byzantine Agreement for Wide-Area Networks". In *12th IEEE International High Assurance Systems Engineering Symposium (HASE'10)*. San Jose USA, 2010

PART III:

CLOUD AUTOMATION

CHAPTER SEVEN

AGENT BASED SERVICES FOR NEGOTIATION, MONITORING AND RECONFIGURATION OF CLOUD RESOURCES

SALVATORE VENTICINQUE

When interoperability and portability of applications across heterogeneous clouds are supported, autonomous services, which can automatically manage discovery, negotiation and monitoring of cloud resources represent an added value for users within the sky computing paradigm. Here we present the design and prototype implementation of an agency that can access, on behalf of the user, the utility market of cloud computing to maintain the best resources configuration that satisfies the application requirements. This system is being designed and developed within the research activities of the FP7 mOSAIC project. It will provide the collection of cloud resources, from different vendors, that continuously meets the requirements of users' applications. According to the available offers, it generates a service-level agreement that represents the result of resource negotiation and booking with available providers. The user is able to delegate to the agency the monitoring of resource utilisation, the necessary checks of the agreement fulfillment and eventually re-negotiations.

1. Introduction

Cloud computing has emerged as the reference paradigm for selling or renting computing resources. Let's imagine the Internet as a huge computer market where customers look for, and get, any kind of computing resources. Computing resources in this context can be hardware, operative systems, development platforms, application servers or complete applications. Virtualisation represents the technology that allows the building of a personalised environment for the user whose configuration and administration is independent from the hardware infrastructure. More virtual machines, which execute on the provider's data centers, allow for the increase of utilisation of the hardware and to isolate the user's environment. In this context common issues are: how to describe one's own requirements, how to buy, how to pay, how to use the booked resources and how to check that what has been bought is compliant with the provider's promises. The business model of cloud computing is "pay per use," meaning that for customers the business model does not introduce a constraint with their provider. Users can sign a new contract when application requirements change or when a new, more convenient offer is available. However this opportunity is limited by the fact that, currently, a provider uses proprietary technologies, which make a transparent migration of users' applications and platforms from one cloud solution to another impossible. This is known as "vendor-lock-in problem." This is a relevant issue that has recently been addressed by the scientific community and is supported by many calls for research projects. In this context, the mOSAIC project proposes a new, enhanced programming paradigm that is able to exploit the cloud computing features for building applications which are able to adapt themselves as much as possible to the available resources and to acquire new ones when needed. mOSAIC will offer an API for development of flexible, scalable, fault tolerant, provider-independent cloud applications, and a software framework for their execution. When interoperability and portability of applications across heterogeneous clouds are supported, autonomous services, which can automatically manage discovery, negotiation and monitoring of cloud resources represent added value for users within the sky computing paradigm (Aversa et al. 2011).

This chapter presents the state of art of research activities on negotiation and other complementary cloud services. Furthermore, it focuses on the agent-based solution proposed within the mOSAIC project and provides details about the latest design choices and developments. Section 2 presents our vision of the Cloud market, with motivation and requirements. Section 3 introduces the mOSAIC project, and related work is discussed in section 4. Section 5 and 6 describe design and prototypal implementation of cloud agency and of provided services. Finally we state our conclusions.

2. The Cloud Market

In order to adopt the sky-computing paradigm, we have to face the cloud as a market of heterogeneous computing resource offered by different vendors. We have to be able to choose the best cloud or to use different clouds together. Being a business model for renting computing resources, we need to address common business problems for when we develop cloud applications. Here we discuss issues of programming from a cloud consumer perspective (Khajeh-Hosseini 2010), through a list of questions from customers put prior to using the cloud.

The first question one should raise is: "How to ask?" (Youseff 2008; Moscato 2011). This means that in order to get a computing resource it should be possible to ask unambiguously, by describing the requirements of final applications which will use those resources. The problem here is not trivial. A first issue concerns the semantic gap between the quality of service which an application running in the cloud must guarantee to the user, and the quality of service which the supplier must provide in order to meet these requirements. A second issue is the semantic representation of the quality and service levels.

A second question is "Where to buy?" It means how to discover available cloud resources, as is done in SOA environments (Vouk 2008). The problem concerns the discovery and the selections of those providers, which offer the necessary resources, and the choice of the technology that allows development and/or deployment of the application. The discovery could take as an input the application requirements, a detailed description of resources at infrastructure level, but also natural language. Furthermore new services and new providers can appear into the cloud scenario and they should be dynamically included. The next question is "How to buy ?" Negotiation represents the process that allows for the choosing of the best solution for the user. Users can interact with providers directly or through a broker (Buyya 2009). Today, cloud providers neither support negotiation protocols nor offer services with defined quality levels. Service levels are usually classified according proprietary labels such as gold, silver and bronze, with different meanings and without warranty. The only property that is usually ensured by many commercial clouds is the maximum down-time in case of failure. The result of a negotiation is a Service Level Agreement (SLA), that is a document where one can find the information about the booked service, its properties, to the method of access and economic parameters of the commercial offer.

After that the user can start to use the resource, but "How to use it?" Here we cannot provide one common answer. The method of using the resource depends on many features: the provider's technology, the type of resource (storage, virtual machine, database, etc), and the cloud level (IAAS, PAAS, SAAS). There are many groups and initiatives which aim at defining the standardisation of cloud services (OCCI, CDMI, etc), but they are still work in progress. Let us suppose that the cloud application is finally running into the cloud. One cannot be sure that they will remain satisfied with their own requirements over time, meaning that the requirements of the application can change dynamically, but also clouds can suffer performance unpredictability (Armbrust et al. 2010), and the current SLA should be updated. For example, the under-utilisation of resources could be too expensive, while the saturation could cause a degradation of application behaviour. On the other hand, the quality of service of the user's application may not meet the requirements because the cloud provider does not grant its promised service levels. Now we need to answer to: “How to check?”. The monitoring that is offered as a service by the provider itself is not enough. The user cannot trust the provider for both the SLA and for its checking. On the other hand some parameters depend by the application and can be monitored only by instrumenting the application itself.

Finally, the last issue we are interested in is “How to change?” the cloud resources, when the monitoring detects critical situations. It deals with the problem of understanding how to act in order to define a new collection of cloud resources which satisfy, at best, the application requirements (Armbrust et al. 2010). Some countermeasures could be: the change of the SLA by adding, resizing or deleting resources; the withdrawal of the SLA and the negotiation of a new one; the choice of a different provider or of a more convenient offer by the same provider.

3. The mOSAIC project

In the current context of the cloud market, mOSAIC is expected to offer the freedom of choice at programming level as well as at the resource level. In order to respond to the above described community needs, the mOSAIC project has been initiated in the frame of FP7-ICT program. It intends to create and promote an open-source cloud application programming interface and a platform targeted for developing multi-cloud oriented applications. The mOSAIC proposal is motivated by the following existing weaknesses of current research and practice in cloud computing:

(1) Lack of common programming model for Cloud-oriented applications
(2) Lack of tools for easy deployment of scalable applications and (multi)-cloud-based service compositions
(3) Lack of standard interfaces for resource virtualisation

(4) Lack of adequate service level agreements and their dynamic negotiation
(5) Platform dependability and non-portability due to different APIs for different types of resources.

To these weaknesses mOSAIC intends to respond by providing the following:

- A language- and platform-agnostic application programming interface for using multi-cloud resources and cloud usage patterns
- An open-source and portable platform for using cloud services based on the proposed API and cloud usage patterns
- A generic agent skeleton for representing various stakeholders, for example vendors and their resources, users of various types, and collection of modules that can be used to adapt agent skeleton to support needed functionalities
- User-centric service level agreements, a cloud ontology and mechanisms for dynamic negotiation of resources based on multi-agent technologies and semantic data processing
- Proof-of-concept applications with emphasis on data intensive applications.

The API refers to a Cloud-based language- and platform-independent application programming interface that extends the existing language- or platform-dependent API capabilities with composite features based on usage patterns. The framework refers to the complete set of scientific and technical solutions provided by the project. Behind the solutions related to the API definition and instantiation, this framework will include especially the semantic-oriented proposals like the cloud ontology, which will allow for semantic representation of resources and direct application of semantically driven information processing. The open-source platform will be a proof-of-the-concept prototype ready to be tested, exploited or extended by its users. It will include instances of the APIs for at least two programming languages and applications tools. Its semantic engine, based on the cloud ontology, will express the application's needs for cloud resources in terms of SLAs and QoS requirements that are the inputs of a negotiation module. This module initiates a bid to the agents representing different cloud resources providers. Cloud resources can vary from software services (including virtual appliances) or data services to hardware services. The software architecture of the mOSAIC framework is shown in figure 7.1.

Fig. 7.1. mOSAIC software architecture

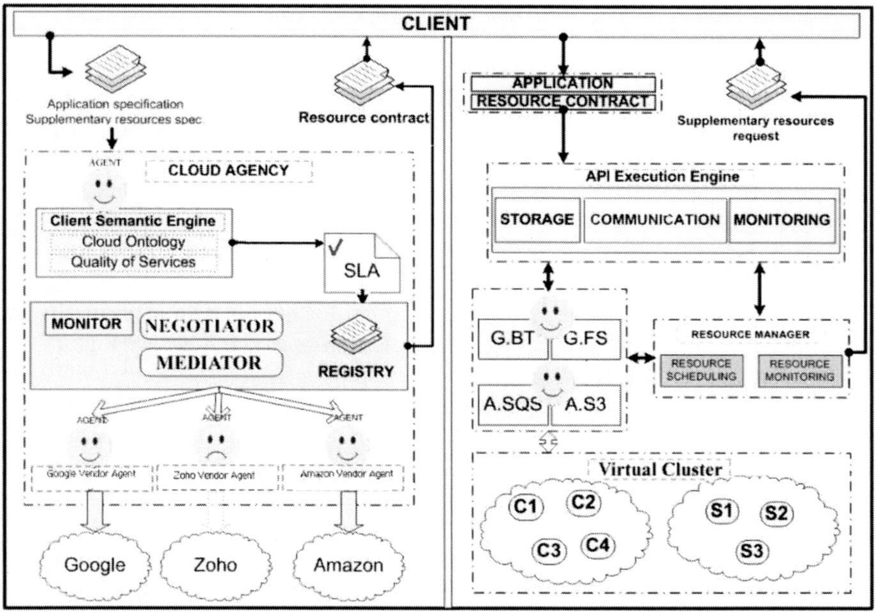

The application requirements, in terms of Cloud resources at infrastructure level necessary to execute the user's application, are submitted to a component called cloud agency whose role is provisioning by negotiation, monitoring and reconfiguration of acquired resources. The agency provides agent-based services to be invoked by a RESTfull or a Web Services interface. In addition, a set of API, part of the mOSAIC framework, will be developed (the first implementation in Java is available at bitbucket.org/mosaic). The outcome of the agency is a Service Level Agreement, that is a contract with cloud providers and all necessary binding information to use the resource. The output of services provided by the cloud agency will be used by the cloud application (CLIENT in the figure) itself and by the mOSAIC runtime for handling scalability and elasticity. The interface of the provisioning subsystem to open source and commercial providers is implemented by vendor agents, who behave as a proxy between the different technology and semantic of providers offer and the one used inside the mOSAIC framework.

4. Related work

The brokering of cloud providers whose offers can meet the requirements of a particular application is a complex issue due to the different business models that are associated with such computing systems. According to Buyya (2009), a market-oriented resource management is needed in order to regulate the supply and demand of cloud resources, providing feedback in terms of economic incentives for both cloud consumers and providers, and promoting QoS-based resource allocation mechanisms that differentiate service requests based on their utility. The current cloud computing technologies offer a limited support for dynamic negotiation of SLAs between participants. There are no mechanisms for automatic allocation of resources to multiple competing requests. Furthermore, current cloud computing technologies are not able to support customer-driven service management based on customer profiles and requested service requirements. It is impossible, according to Buyya, to derive appropriate market-based resource management strategies that encompass both customer-driven service management and computational risk management to sustain SLA-oriented resource allocation.

An attempt to define several QoS metrics is presented in Cao (2009): response time, availability, reliability, cost and reputation are considered. A reference of SLA model is provided in Sim (2010) where SLA objectives (SLOs) are used to compose an SLA. A number of service levels and performance metrics for each resource results in multiple SLOs for every service. The work presented in Kertesz (2009) represents a first proposal to combine SLA-based resource negotiations with virtualised resources in terms of on-demand service provision. The architecture description focuses on three topics: agreement negotiation, service brokering and deployment using virtualisation. It involves multiple brokers. A cloud multi-agent management architecture is proposed in Bu-Qing Cao (2009). It includes requestor agents, QoS agent, provider agent and agent manager. In particular the QoS agent is dedicated to the submission of QoS information about services.

An agent based architecture has been proposed in You (2009). It is simpler than that described in this paper and consists of only three parts: consumer agent delegated by his owner to obtain a maximal benefit for him; resource agent delegated by the resource provider to publish the prices and to adjust them according to the relationship of supply and demand in the market system; and the market economy mechanism responsible for balancing the resource supply and the demand. New SLA-oriented resource management strategies must be designed for clouds in order to provide personalised attention to customers. Service requirements of users can change over time, due to continuing changes in business operations and operating environment, and thus may require amendments of original service requests. Effects due to concurrent negotiation activities between broker and provider agents in multiple cloud resource markets have to be considered. The negotiation between broker and provider agents influence the negotiation outcomes of broker and consumer agents in a cloud service market. The implementation presented in Sim (2010) is very interesting from this point of view because it supports dynamic negotiations. Preliminary investigations on related topics have been presented in Venticinque (2010 and 2011).

Among research projects working in this context, very interesting results come from the RESERVOIR project (www.reservoir-fp7.eu). The Resources and Services Virtualization without Barriers Project is an European Union FP7 funded Grid Computing project. RESERVOIR research aims to harness the power of virtualisation and grid technologies across administrative domains, and to provide a foundation for a cost-competitive service-based online economy where resources and services are transparently and flexibly provisioned and managed like utilities. The RESERVOIR project is intended to introduce a technological infrastructure for the set-up and deployment of services on demand, and competitive costs, across disparate administrative domains, while assuring quality of service. A Management Layer provides dynamic deployment and relocation of Virtual Environments (VE) on underlying physical resources, based on quality of service requirements from a Service Level Agreement across administrative domains, such as multiple-service sites. A Service Management Layer will provide the interface for requirements from the business world, including support for billing for services used, composition of the definition of the service required, and the monitoring of SLA compliance. Many outcomes of the RESERVOIR project are integrated into the OpenNebula solution for the management of virtualised data centers. This solution works at IaaS level of cloud, but not at PaaS as mOSAIC.

SLA@SOI (http://sla-at-soi.eu) is another European project which aims (together with other relevant goals) at offering an open source based SLA management framework that will provide benefits of predictability, transparency and automation in an arbitrary service-oriented infrastructure. SLA@SOI results are extremely interesting and offer a clear starting basis for the SLA management in complex architectures. SLA@SOI offer solutions to design cloud services with multi-level and multi-provider SLAs . The main target of the project is Cloud Provider enrichment for offering a solution that can be integrated with Cloud technologies (like Open Nebula) trough an SLA-based approach.

SLA management has been investigated by the IRMOS project (www.irmosproject.eu) to develop “real-time” interaction between people and applications over a service oriented infrastructure (SOI), where processing, storage and networking need to be combined and delivered with guaranteed levels of service. In this scenario it needs real time technologies and SOIs that are able to deliver the right power on demand. Relevant application domains are film post-production, virtual augmented reality for engineering design and collaborative e-learning. IRMOS outcomes consist of PaaS management models, tools and services that allow application services providers to match cloud resourcing policies to application workloads defined in machine readable Service Level Agreements, considering real-time Quality of Service (QoS) requirements. Its approach assumes uncertainty in provisioning decisions and manages this uncertainty through the use of stochastic system models within runtime QoS management processes.

5. The Cloud Agency

The provisioning services of the mOSAIC framework are in charge to implement a set of functionalities for resources negotiation and management. These facilities can be used by the mOSAIC Runtime, if the developer uses the mOSAIC SDK to implement his applications and wants to delegate the resource management. Otherwise these services can be invoked independently by the user application itself, if it is able to use the provisioning facilities in a custom way. Cloud agency is in charge of provisioning of cloud resources for mOSAIC applications. Negotiation, monitoring and reconfiguration at the cloud Infrastructure Level are the three agent based services available for cloud developers and for the mOSAIC framework. Before its deployment, cloud agency executes on the desktop of the programmer. When computing resources for running the mOSAIC applications are available, cloud agency is running in the cloud. Cloud agency is composed of a number of containers which are dynamically spawned on each new booked computing resource. It can reserve new resources, or release the active ones, and reconfigure itself in the meantime. Two usage scenarios have been conceived:

- **Cloud agency as an independent component**. As an independent component cloud agency can be used just to book cloud resources and eventually to monitor and reconfigure them also without programming with the mOSAIC SDK. Cloud agency will be able to scale up, scale down resources or change the providers in an autonomic way. Otherwise users can access a graphical interface or they can invoke cloud agency services from their applications using a RESTfull interface.

- **Cloud agency as a component of the mOSAIC framework**. As a component of the mOSAIC framework cloud agency offers its services to the core components of the framework (for example reconfiguration at application level) and to the mOSAIC applications. A connector will be developed to use services within the cloudlets. Connector and cloudlet are building blocks for developing mOSAIC applications.

Negotiation, monitoring and reconfiguration of the cloud at resource level are the three services which are provided by the Agency. Services are invoked and used by the cloud agency API as follows.

5.1 Cloud Agency Interface and APIs Design

Cloud agency interface provides an asynchronous APIs that is compliant with the event-driven mOSAIC programming model. To address this issue use cases are designed in terms of service requests, events and callbacks. Access to cloud agency services will be enabled by HTTP RESTfull interface. This is the model we will use to design cloud agency APIs within the mOSAIC project:

- **Service Requests.** They are used to ask the cloud agency for something to be executed. For example, to start a negotiation, to accept or to refuse an SLA, to change a policy, etc. Requests, as any other actions, are not-blocking. It means that execution is started remotely, but the client can continue to run. On the other hand, requests will generate future events, which have to be handled by the requester.
- **CallBacks.** They are used to receive and handle events which follow previous requests. Programmers must implement callbacks in their applications.
- **Queries.** They are used to get information. For example clients can ask for reading an SLA, the status of a negotiation, to get the list of vendors, or the list of resources. Queries are synchronous, they return immediately the response if it is available, and otherwise an exception.

As an example, some actions are listed in table 7.1 below.

Table 7.1. Examples of Cloud-Agency API

N.	Action	Actor	Type	Triggered Event	Trigger Event
r1	startCFP	Client	Req	ProposalAvailable	
q1	listVendors	Client	Query		
q2	listSLAs	Client	Query		
q3	getSLA	Client	Query		
r2	acceptProposal	Client	Req	SLAAvailable	
cb1	Proposal Available	Client	CB		ProposalAvailable
r3	Query-REF	Client	Req	INFORM	
cb2	Inform-REF	Negotiator	CB		INFORM

The client uses the request "r1" to start a negotiation. It can be done by using the mOSAIC API, or by a CA GUI Interface. Concurrent negotiations are not yet supported. When a proposal is available the Agency returns a "ProposalAvailable event" to the client that will be intercepted and managed by the "ProposalAvailable callback." The query "q1" is used to get the list of available vendor agents. They are searched into the directory facilitator (DF), a yellow page service for agents platforms. Some parameters can be used for an advanced search. With the query "q2," the client gets the list of SLAs. Parameters can be used to filter the active ones, those closed in a certain period, and those that involve particular providers. SLAs are stored internally. Query "q3" allows the reading of a specific SLA whose identification string is provided as a parameter. Request "r2" is used to accept a proposal. The request is asynchronous in the sense that the client continues its execution while the agency contacts all the providers involved, to wait for the acknowledgement, the parameters for binding and the set-up time. Furthermore, the transaction could fail because of different reasons, such as a timeout. Failure and success of this action are notified by events. "ProposalAvailable" is a CallBack to be implemented by the client, started when an event with the same name is received from the cloud agency. The programmer needs at least to evaluate the proposal and to accept or refuse it. Request "r3" and callback "cb2" are useful to exchange any kind of general information which are not strictly defined by the already designed functionalities. Using these actions the protocol can be extended and the kind of action can be specified into the body of the message.

5.2 HTTP rendering of Cloud Agency APIs

The proposal extends the OCCI results and adopts the same rendering for common functionalities, such as starting a virtual machine, or to ask for available computing resources. GET/POST messages are used to start requests, to submit query or notify events. Objects and resources are identified by abstract paths and files, which are specified by special headers added to the HTTP message. As an example, let us consider some examples, which implement the APIs described above.

The **startCFP** request allows the user to send a call for proposal to the cloud agency. A POST message, embedding an SLA a template that must be formatted according to the WSAgreement model, is sent to the agency. An example of rendering is shown in fig. 7.2 below. In particular, the reply-to parameter is used to allow the asynchronous response. All the events, which will be raised by the request, will be notified at the specified URL.

When the request has been received, the cloud agency starts the corresponding action and returns an acknowledge with the identification string for that transaction, to be used for the following interactions. The acknowledge rendering is shown in fig. 7.3 below. In particular, the location parameter of the HTTP header allows the client to identify that transaction.

The ProposalAvailable event is notified to the client by the cloud agency using a POST message again. As shown in fig. 7.2 below, the message is sent to the CallBack handler that must recognise the event and start the necessary action for accepting or refusing the proposal. The message notifies the event, but does not attach the proposal that can be read using a cloud agency query. The location header is the necessary reference to get that proposal. In the example, for the correspondent CFP, we have only one proposal that is identified by the same string and by a different path.

Fig. 7.2. REST message for starting Call for Proposal

```
POST / HTTP/1.1
User-Agent: curl/7.21.0 (x86_64-pc-linux-gnu) libcurl/7.21.0
OpenSSL/0.9.8o zlib/1.2.3.4
Host: localhost:8080
Accept: */*
Cookie:
pyocci_user=Zm9v|1291753962|7011a4821179ff98ea96d4b44fade0512b1
ffc52
Content-Type: text/occi
Reply-To: localhost:8080/CallBackHandler
Category: cfp; scheme="http://schemas.ogf.org/occi/provisioning#";
class="mixin";
Content-Length: …
<ws:Template ws:TemplateId="T-1"
xmlns:ws="http://schemas.ggf.org/graap/2007/03/ws–agreement">
 <ws:Name>Test-1</ws:Name>
 <ws:Context> … </ws:Context>
 <ws:Terms> …</ws:Terms>
 <ws:CreationConstraints/>
</ws:Template>
```

Fig. 7.3. REST response of Call for Proposal

```
HTTP/1.1 200 OK
Content-Length: 2
Content-Type: text/html; charset=UTF-8
Location: /users/foo/cfp/b9ff813e-fee5-4a9d-b839-673f39746096
Server: CloudAgency OCCI/1.1
```

The client responds by a general "HTTP/1.1 200 OK" message to acknowledge the receive of the event (fig. 7.4 below). Finally, to read the content of the proposal a simple "HTTP GET" is used. It needs to specify the location from the previous message:

GET /users/foo/proposals/b9ff813e-fee5-4a9d-b839-673f39746096 **HTTP/1.1**

Fig. 7.4. HTTP notification of Proposal Available

```
POST / HTTP/1.1
User-Agent: CloudAgency OCCI/1.1
Host: localhost:8080/CallBackHAndler
Cookie:
pyocci_user=Zm9v|1291753962|7011a4821179ff98ea96d4b44fade0512b1
ffc52
Event: ProposalAvailable
Location: /users/foo/proposals/b9ff813e-fee5-4a9d-b839-673f39746096
Accept: OCCI/1.1
```

Fig. 7-5. REST response to a query for reading the Proposal


```
HTTP/1.1 200 OK
Content-Type:text/html
Content-Length:...
\n\r
<ws:Template ws:TemplateId="t1"
xmlns:ws="http://schemas.ggf.org/graap/2007/03/ws–agreement">
 <ws:Name>Test1</ws:Name>
 <ws:Context>
 <ws:ServiceProvider>google</ws:ServiceProvider>
 <ws:TemplateId>t1</ws:TemplateId>
 <ws:TemplateName>Test1</ws:TemplateName>
 </ws:Context>
 <ws:Terms>
 <ws:All>
  <ws:ServiceDescriptionTerm ws:Name="google_storage01"
ws:ServiceName="SET VARIABLE">
  <mod:Storage xmlns:mod="http://occi–wg.org/model">
   <title>ANY</title>
   <size>1GB</size>
   <summary>ANY</summary>
   <writetime>1MBs</writetime>
  </mod:Storage>
  </ws:ServiceDescriptionTerm>
 </ws:All>
 </ws:Terms>
 <ws:CreationConstraints/>
</ws:Template>
```

This time, the response from the cloud agency contains a body that describes the template previously sent, as shown in fig. 7.2 above, but where each parameter has been filled according to the result of the negotiation. In fig. 7.5, a possible response is shown. In particular the attached XML describes an SLA for a storage. The provider is Google. The storage size is 1 GB and the granted write time is 1MB/sec.

At client side, the user is provided with the GUI shown in fig. 7.6, which allows the starting of requests and queries presented before. It notifies events and visualises the incoming messages.

Fig. 7.6. Cloud-Agency GUI

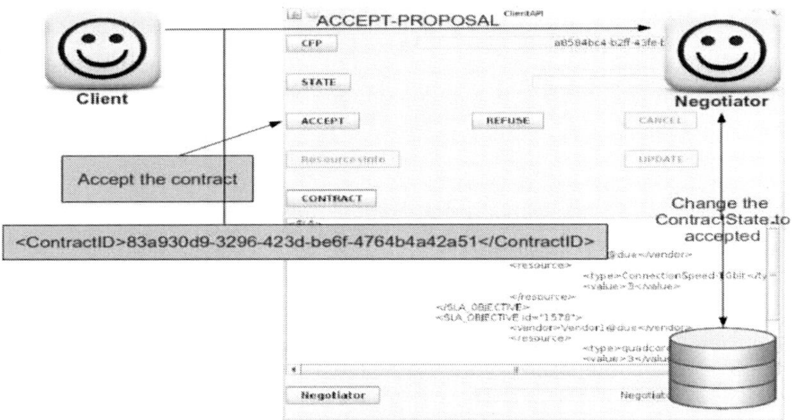

6. Software Architecture and Facilities

In fig. 7.7, the software architecture of the cloud agency is shown. The depicted components implement three services—negotiation, monitoring and autonomic configuration—using a programming paradigm based on distributed agents. The Negotiator takes care of starting a call for proposal after the correspondent user's requests. It contacts Vendor Agents, which can talk to the supported cloud vendors. The Archiver collects information about performance and availability of active resources by distributed bench-markers and agent meters, and computes the relevant statistics. A rules based Observer monitors the collected data and statistics in order to detect and log critical situations. These events are notified using a Bus Event that can be used either by external processes and by the cloud agency itself. Another Observer subscribes to the Bus Event to receive internal and external events. It uses a prolog engine to process the history of events and a set of rules to deliberate the reactions to be applied. Some examples of decisions on reconfiguration strategies are: a new negotiation, the termination of a contract, a change of current agreements. Details are provided in the following subsection.

Being implemented as a MAS (Multi Agent System), the internal workflows of the agency are intrinsically asynchronous, because agents react on the occurrence of a message receive. At the same time, they launch new events by sending messages to other agents.

Fig. 7.7. Software architecture

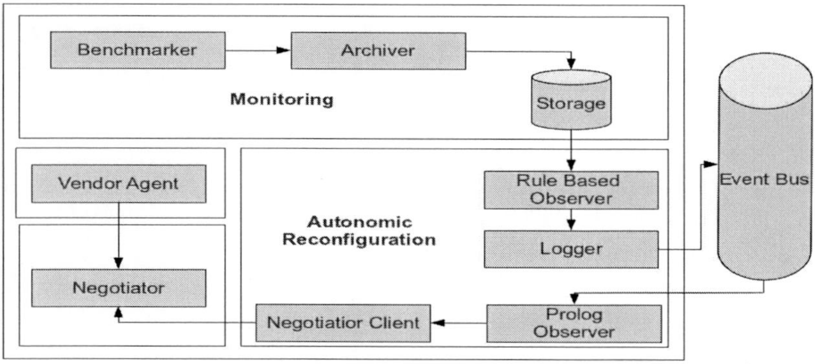

6.1 The Negotiation Service

This service is implemented by distributed and specialised agents, which interact using standard protocols. A Client Agent acts as access point. It receives the user's requests and cooperates with the Negotiator in order to book cloud services and resources with the requested quality levels. The Client here represents an agent that acts as a proxy between the REST/OCCI protocol and the internal Agent Communication Language (ACL). The Negotiator Agent receives from the Client an SLA template (a call for proposal) and splits it into different objectives. It receives different proposals from one or more vendors. It evaluates the proposals and builds the final SLA. The Mediator is a broker that, for each call for proposal, which is coming from the Negotiator Agents, searches in the DF the available vendors. It contacts each Provider Agent and requests a bid for the needed resources. The Vendor Agent works as a wrapper, or a driver. It translates the specific provider protocol into the uniform protocol used inside the mOSAIC framework.

Fig. 7.8. Negotiation protocol

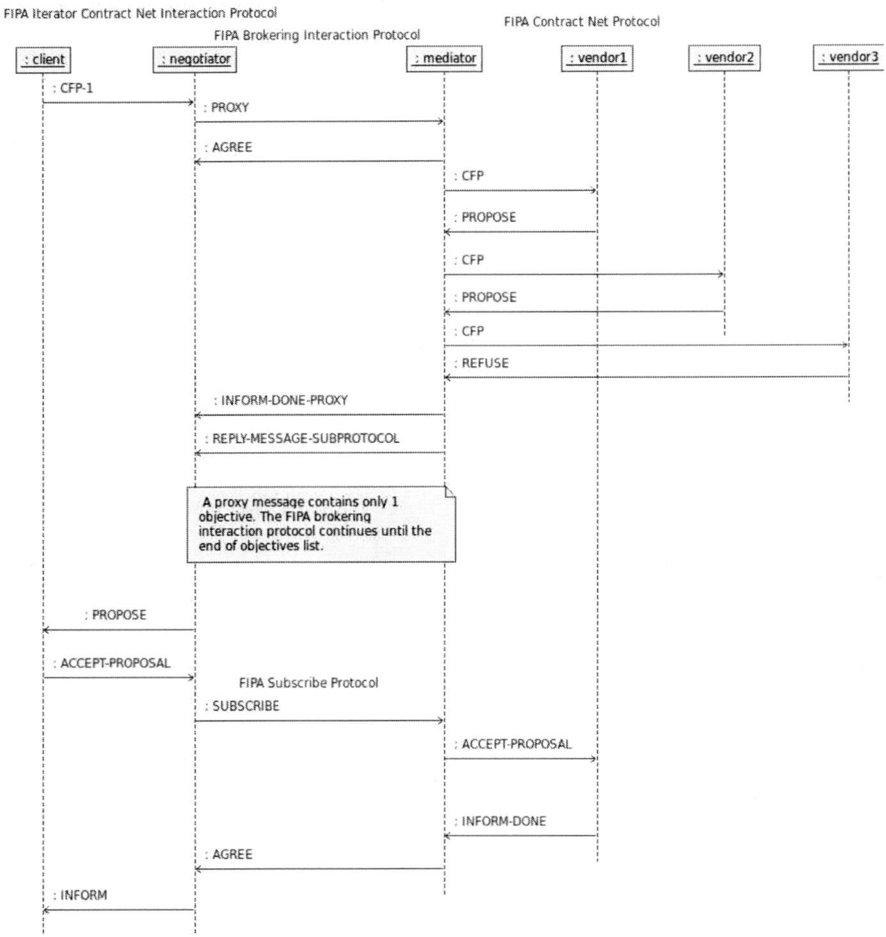

The cooperation among agents has been designed adopting well defined Agents Interaction Protocols of the FIPA standard (Grama 2000). The Client Agent implements a standard FIPA Contract Net Interaction Protocol with the Negotiator Agent. It submits a call for proposal using an SLA Template with the application requirements and gets the SLA proposal. The Mediator implements the standard FIPA Brokering Interaction protocol. The sub-protocol used to talk with vendors is again the standard FIPA Contract Net interaction protocol. A successful negotiation that involves three vendors is shown in fig. 7.8. In the provided example we suppose that the template is split by the Negotiator into different objectives. The Mediator for each objective contacts one vendor. The offers from the different vendors are combined or filtered by the Negotiator again that notifies to the client the resulting proposal.

The behaviour of each agent has been designed by an automata. In fig. 7.9 below the behaviour of the Negotiator is shown. At start-up, the agent is waiting for a CFP. A request for starting a new negotiation must contain an SLA template. If the message cannot be read, the automata fails and starts waiting again after a reset message. If the request can be handled the template is parsed and a request for searching available providers is sent to the Mediator. If the Mediator accepts the message the Negotiator waits again until all providers have been invited to propose and offer. The Mediator informs about how many proposals are expected. Each time a new proposal has been received the Negotiator tries to build an SLA proposal. When a SLA proposal is ready the client will be informed. The SLA can be accepted or refused. If it is accepted, a new waiting state blocks the Negotiator until all the resources are booked and available.

Fig.7.9. Asynchronous behaviour of the Negotiator

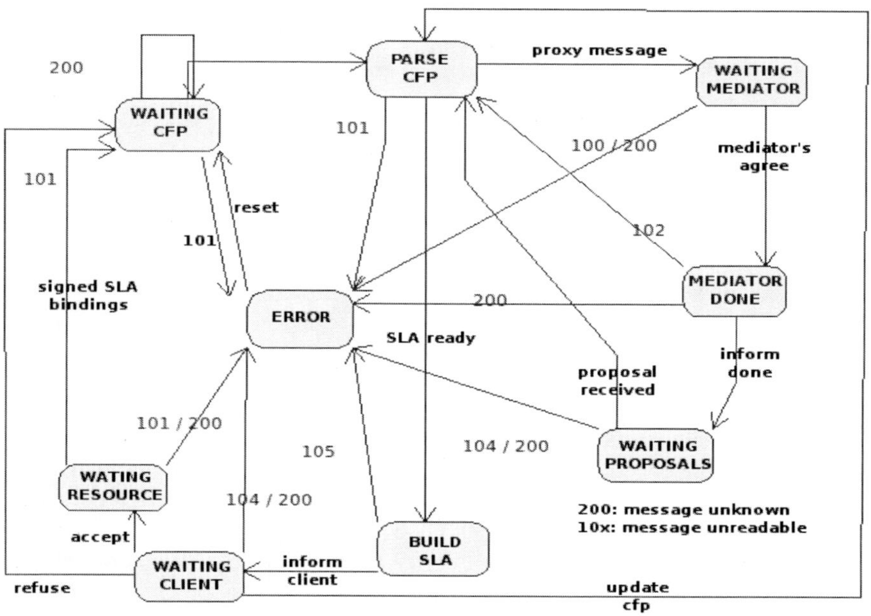

6.2 Monitoring

The Monitoring service allows for the collection of information and measures which can help to detect critical situations of cloud. The monitoring activity is never for free in terms of overhead, so it is necessary to configure the service in order to collect only the necessary information. For this purpose the CA provides the client with the list of parameters, which can be monitored for a given resource type.

Fig. 7.10. Archiver Agent for monitoring

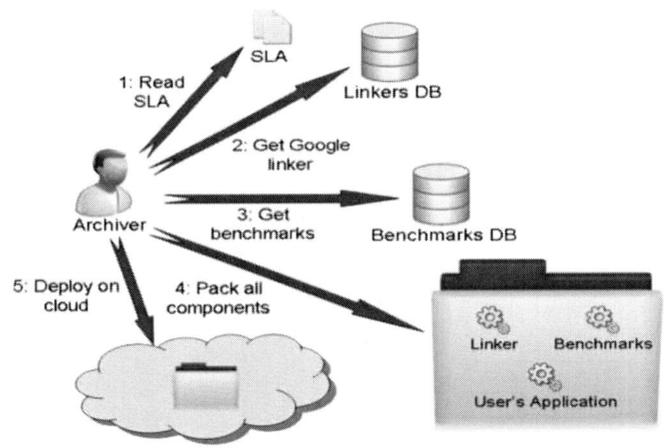

The client finds the possible monitoring parameters that can be used to compose metrics and can choose what metrics to be added/removed/ retrieved. A metric is a statistical function (e.g. max, min, average, sum) of one monitoring parameter of one resource type over a certain period. CA notifies to the client the values of those metrics for which the client requested notifications. In fig. 7.10 above, Benchmarking Agents collect measures about the QOS parameters to be monitored. They can perform measures while executing into the cloud, within each virtual machine. The Archiver Agent maintains statistics about the mOSAIC system. It uses Benchmarking Agents to measure QOS parameters and to store their values. The parameter to be measured could be a QoS level specified into the SLA or any information that is relevant to understand the behaviour of the application. Using the SLA, the Archiver (see fig. 7.10 above) is able to select the right meter, according to the parameter to be measured and the kind of resource, and the library (linker) that is compliant with the technology of the specific provider. The agent is deployed into the cloud at the right necessary destination.

In order to detect and to be notified about critical situations, the user can specify a set of rules: by a set of API, or by the GUI shown in fig. 7.11 below. Using our demonstrator, the user can select an aggregation metric (average, max, min, sum) for each parameter. The computed value can be compared using relational operators with a fixed value or with the level granted into the SLA. We have to specify when the rule should be checked by defining a time period or an event. Events are triggered by incoming messages. When the check fails it is possible (a) to activate an idle rule; (b) to pause an active rule; or (c) to notify the event to all agents or to specific ones. The defined rule can start active or idle mode.

Fig. 7.11. GUI for definition of monitoring rules

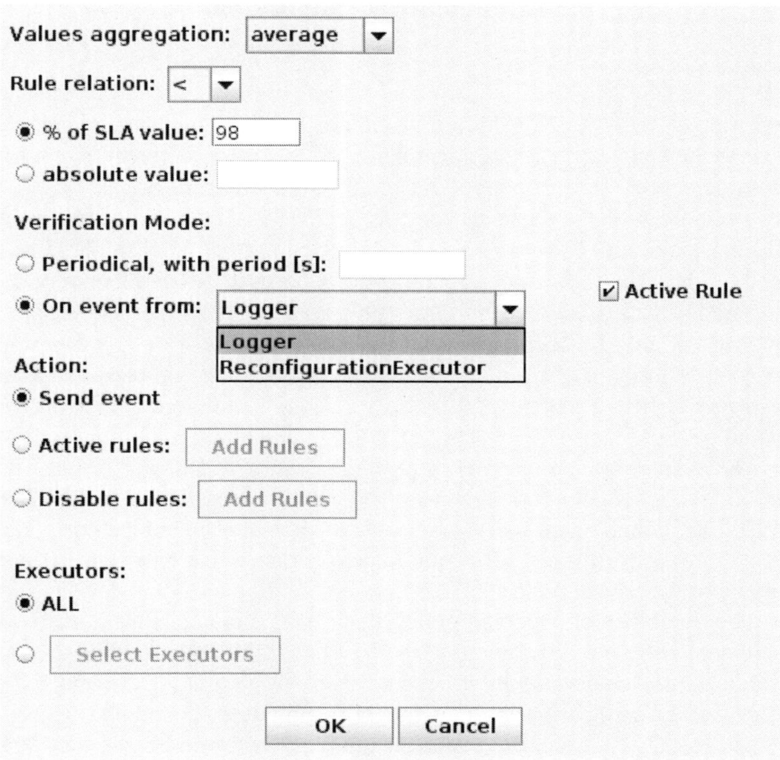

Some examples of rules are shown in fig. 7.12 below. The window allows for selection from the active SLA a provider, a resource among the ones booked on that provider, and the parameter to be monitored. After that, it is possible to add a new rule for that parameter.

Fig. 7.12. Window to add, remove, watch rules of the monitoring profile

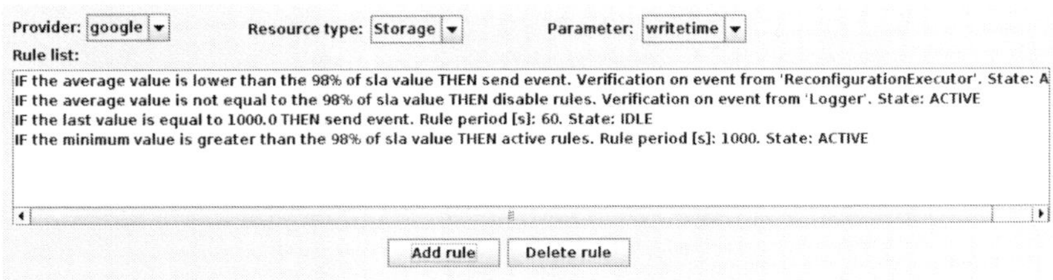

6.3 Autonomic reconfiguration

Cloud reconfiguration can be useful for many reasons. A client could choose to change provider when it does not fulfill the SLA conditions. The detection of this situation currently cannot be used to claim penalties, because the state of the art commercial providers offer different levels for the same service (bronze, silver and gold, all at different prices) without any guarantees of the performance levels. As we said before, they grant only the maximum downtime for the provided service in case of failure. On the other hand, the reconfiguration can be used to avoid or recover undesired situations. When an under-utilisation of resources is detected, the user can look for a cheaper offer and can switch to it or to a different provider. Saturation of resources could cause a degradation of application performances and a more expensive service could be chosen to recover the desired quality.

The event can be a resource failure or a metric evaluation result. The conditions are usually comparisons of metric results with thresholds. The actions are either client alerts or resource allocations/deallocations. The client can retrieve the list of available policies or define new ones. It can activate or deactivate policies.

The capability of checking policies and of taking decisions have been implemented by multiple agents. Fig. 7.13 below distinguishes Observers and Executors. Observers subscribe to an event-bus in order to communicate with external applications that can send information or activate triggers asynchronously. Observers evaluate policies, take decisions and ask Executors to perform specific actions.

Fig. 7.13. Agents based model for autonomic reconfiguration

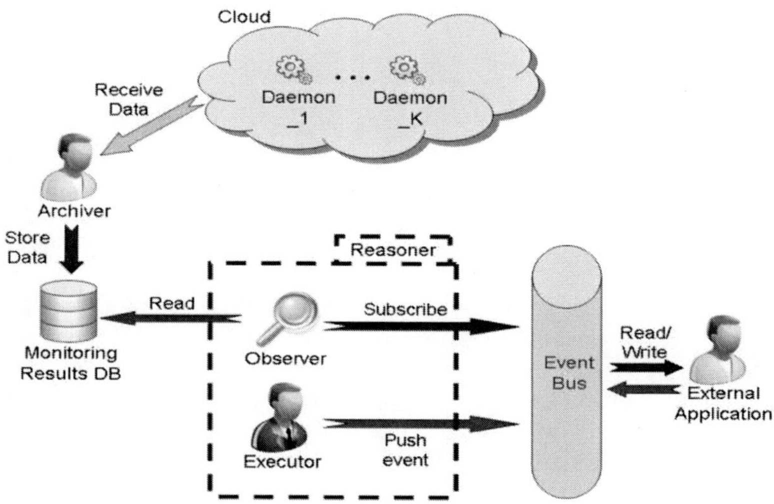

In our demonstrator we developed four agents. A first Observer monitors the metrics collected by the Archiver and checks the rules as described in the previous sub-section. When a check fails, the Observer asks an Executor (Logger) to push the event by the Event-Bus. A second Observer uses a Prolog engine to process a set of rules which can be defined by the user. The second Observer receives the internal and external events by the Event-Bus and updates its knowledge base. For each received message it runs the reasoner and gets the actions to be performed, if any are necessary.

In fig. 7.14 below, a set of rules are shown. Lines 1 and 14 tell that each event must be dynamically added to the knowledge base. Lines 2 to 5 specify what events will trigger saturation condition. Lines 6 to 9 specify what events will trigger the slow condition. At lines 10 and 11 two available actions are declared—the first asks to increase the storage size, the second adds a virtual machine. Lines 12 and 13 specify which triggers activate the two actions and define which parameter should be passed.

Fig. 7.14. Prolog profile for reconfiguration

```
1. :- dynamic (addEvent/6).

2. saturation (ID) :- event (storage, ID, free_space, average, lower, X),
   X < 5 .
3. saturation (ID) :- event (storage, ID, free_space, average,
   lower_equal, X), X =< 5 .
4. saturation (ID) :- event (storage, ID, free_space, last, lower, X), X <
   5 .
5. saturation (ID) :- event (storage, ID, free_space, last, lower_equal,
   X), X =< 5 .
6. slow (ID) :- event (storage, ID, free_space, average, lower, X), X < 5
   ,event(compute, ID, cpu_usage, average, greater,Y), Y > 95 .
7. slow (ID) :- event (storage, ID, free_space, average, lower, X), X < 5,
   event(compute, ID, cpu_usage, average, greater_equal,Y), Y > 95 .
8. slow (ID) :- event (storage, ID, free_space, average, lower_equal, X),
   X < 5, event(compute, ID, cpu_usage, average, greater,Y), Y > 95 .
9. slow (ID) :- event (storage, ID, free_space, average, lower_equal, X),
   X < 5, event(compute, ID, cpu_usage, average, greater_equal,Y), Y
   > 95 .

10. action (saturation, update, storage) .
11. action (saturation, add_resource, compute) .

   % program interface
12. performAction (ACTION, RESOURCE, ID) :- action (saturation,
   ACTION, RESOURCE), saturation (ID) .
13. performAction (ACTION, RESOURCE, ID) :- action (slow,
   ACTION, RESOURCE), slow (ID) .

14. addEvent (RESOURCE, ID, PARAETER, AGGREGATION,
   RELATION, PERCENTAGE_VALUE) :- asserta (event
   (RESOURCE, ID, PARAMETER, AGGREGATION, RELATION,
   PERCENTAGE_VALUE)) .
```

7. Conclusion

In this chapter we described the requirements, the design and the prototype implementation of cloud agency, an agent based software that provides provisioning capability to cloud. Cloud agency has been conceived to be used both as a stand-alone software and as a component of the mOSAIC framework. The agency exposes negotiation, monitoring and autonomic reconfiguration services. Asynchronous APIs have been designed and provided by a REST interface, which follows the OCCI reference model.

To execute services, agents communicate using standard FIPA messages and according to FIPA interaction protocols. A demonstrator has been implemented using the Jade platform. Detection of critical situations and autonomic reconfiguration of cloud resources are based on Policies. Policies are event-condition-action rules that allow to design autonomic behaviour in a flexible and dynamic way. A Prolog prototype has been developed. Future developments aim to engineer the complete cloud agency also by implementing and testing effective algorithms for negotiation, and to define complete profiles for monitoring and reconfiguration in real case studies. Components of the mOSAIC framework that allows to automatically integrate the agent services into the cloud applications will be developed.

Acknowledgment

The work described in this chapter has been supported by the FP7-ICT-2009-5-256910 (mOSAIC) EU project.

References

Armbrust, Michael et al. "A View of Cloud Computing." *Communication of ACM* 53 (4) (2010): 50–58.

Aversa, R., Venticinque, S., Di Martino, B. and Petcu, D. "Agent based Cloud Provisioning and Management, Design and Prototypal Implementation". *1st International Conference on Cloud Computing and Services Science (CLOSER2011).* SciTePress. 184-191, 2011.

Buyya Rajkumar et al. "Cloud Computing and Emerging IT Platforms: Vision, Hype, and Reality for Delivering Computing as the 5th Utility." *Future Generation Computer Systems* 25 (6) (2009): 599–616.

Cao, Bu-Qing, Li, Bing and Xia, Qi-Ming. "A Service-Oriented Qos-Assured and Multi-Agent Cloud Computing Architecture." CloudCom, New York: Springer, 2009.

Grama, A., Kumar, V. and Sameh, A., "Foundation for intelligent physical agents", 2000. http://www.fipa.org (accessed September 15, 2011)

Kertesz, A., Kecskemeti, G. and Brandic, I. "An SLA-based resource virtualization approach for on-demand service provision." 3rd International Workshop on Virtualization Technologies in Distributed Computing. 2009.

Khajeh-Hosseini, A., Sommerville, I. and Sriram, I. "Research challenges for enterprise cloud computing." In ACM Symposium on Cloud Computing. ACM CoRR, 2010.

Moscato, F. et al. "An Ontology for the Cloud in mOSAIC." *Cloud Computing: Methodology, System, and Applications*. Sydney: CRC Press, 2011.

Sim and Kwang. "Towards Complex Negotiation for Cloud Economy." In *Advances in Grid and Pervasive Computing*. New York: Springer, 2010.

Venticinque, S., Aversa, R., Di Martino, B., Rak, M. and Petcu, D. "A Cloud Agency for SLA Negotiation and Management". In *Euro-Par 2010, Parallel Processing Workshops,* 547–554. New York: Springer. 2011.

Venticinque, S., Aversa, R., Di Martino, B. and Rak, M. "Cloud Agency: A Mobile Agent Based Cloud System." In *International Conference on Intelligent, Complex and Intensive Systems*. IEEE Computer Society, 2010.

Vouk, M. A. "Cloud Computing—Issues, Research and Implementations." *Journal of Computing and Information Technology* 16 (4) (2008): 235–246

You, X., Xu, X., Wan, J. and Yu, D. "RAS-M: Resource Allocation Strategy Based on Market Mechanism in Cloud Computing." Fourth ChinaGrid Annual Conference 2009.

Youseff, L., Butrico, M. and Da Silva, D. "Towards a Unified Ontology of Cloud Computing", Grid Computing Environments Workshop (GCE08), 2008.

CHAPTER EIGHT

DESIGN FOR SELF-ADAPTATION IN SERVICE-ORIENTED SYSTEMS IN THE CLOUD

ANTONIO BUCCHIARONE, CINZIA CAPPIELLO,
ELISABETTA DI NITTO, SERGEI GORLATCH,
DOMINIQUE MEILÄNDER AND ANDREAS METZGER

Service-oriented systems are able to offer complex and flexible functionalities in widely distributed environments by composing different types of services. These systems have to be adaptable to unforeseen changes in the functionality offered by component services and to their unavailability or decreasing performances. Furthermore, when systems are made available to a high number of potential users, they should also be able to dynamically adapt to the current context of use as well as to specific requirements and needs of the specific users. In order to address these issues, mechanisms that enable adaptation should be introduced in the life-cycle of systems, both in the design and the runtime phases.

In this chapter we will go through the life-cycle of a service-oriented system highlighting those activities needed to support adaptation. The adaptation activities can be performed at various layers of the service-oriented system. In particular, they can concern the layer where services are composed together or the layer of the executing infrastructure, typically, a cloud system. To exemplify the various steps and activities, we use an example from the domain of real-time online interactive applications.

1. Introduction

Modern software technology has enabled us to build software systems with a high degree of flexibility. The most important development in this direction is the concept of service and Service-Oriented Architecture (SOA) (Josuttis 2007). A service-oriented system is built by composing software services (and is thus also called service composition or composed service in the literature).

Software services achieve a high degree of flexibility by separating ownership, maintenance and operation from the use of the software. Service users do not need to acquire, deploy and run the individual piece of software, because they can access the functionality of that software remotely through the service's interface. Ownership, maintenance and operation of the software remain with the service provider (Di Nitto et al. 2008).

One key principle of SOA is "loose coupling," which means that a service only makes weak assumptions about its interactions with other services; for example, instead of services being tightly coupled by means of a common data model, only simple data types are used. This principle allows a service to be (re)used in many different service compositions. Another key principle of SOA is "late binding." This implies that many services will be discovered and composed into a service-oriented system only at run-time. Service-oriented systems thus provide their functionality rather dynamically and in a loose fashion.

As recent studies reveal (Zheng et al. 2010), such distributed applications are extremely fragile, due to the execution of uncontrollable third-party services. Unexpected changes of third-party services or unpredicted network latencies, for example, can cause failures which negatively impact on the timely execution of the applications. To prevent such failures, self-adaptation capabilities are needed.

Self-adaptation is usually referred to as the ability of a system to autonomously address changes during its operation. The activities performed during adaptation typically follow the steps of the MAPE loop, which stands for Monitor-Analyze-Plan-Execute (Salehie et al. 2009).

Service-oriented systems operate on different layers to implement their functionality and to provide the expected quality. In a simplified view, we can identify two main layers; one of the computational and storage resources used for executing the system (often called the "infrastructure layer"), and the "composition layer," in which the services offered by the system are made available and composed together. The infrastructure layer can be organised in various ways and encompasses the boundaries of various organisations. In the case when services are offered by various third parties we can have different cloud systems, either private or public, making available their resources to the system.

In this two-layer setting, self-adaptation can happen both at the infrastructure and at the service composition layer, depending on the nature of the situation to be accommodated. If, for instance, a failure of a computational resource occurs at the infrastructure layer it is quite natural to design such a layer so that it is able to replace the failing resource with another. Similarly, if the failure concerns a specific service operating as part of the system, at the service composition layer, it is possible to replace the failing service with another without making the infrastructure layer aware of this change.

While in many cases it is possible to isolate self-adaptation in a single layer, there are cases in which this is not possible. In particular, adaptations on different layers might impact on each other and may even be conflicting. As an example, if an adaptation on the infrastructure layer leads to the migration of a service from one cloud to another, this could be conflicting with the decision on the service composition layer to choose a third-party service instead of the "own" service implementation deployed in the cloud. In addition, there may be cases in which the joint adaptation on both layers provides opportunities to address problems, which isolated applications on single layers will not be able to remedy.

The need to scrutinise such cross-layer adaptations has been elaborated in S-Cube, the EU Network of Excellence on Software Services and Systems (Papazoglou et al. 2010). Additionally, a recent survey amongst professionals from the Future Internet community, which has been performed during the Future Internet Assembly (FIA 2011) in Budapest, has confirmed the importance of such cross-layer adaptations (Metzger et al. 2011).

In this chapter we focus on two main aspects, that is, the kinds of changes that trigger self-adaptation in a service-oriented system and the strategies that can be adopted to deal with adaptation. We do so by highlighting the fact that some adaptation can be performed at the infrastructure layer while others have to be handled at the composition layer, and we will scrutinise design activities and decisions for applications that will be deployed in the cloud as the infrastructural target. We also provide a preliminary contribution to the systematic understanding of adaptation across layers.

To exemplify the approach presented in the chapter we refer to a specific system example focusing on the domain of Real-Time Online Interactive Applications (ROIA). This system has been developed by exploiting a service-oriented approach and has been featured with a software component that deploys and manages it on the cloud. Among other things, this component is particularly interesting for us as it is able to execute some cross-layer adaptations.

2. Kinds of self-adaptation needs

Self-adaptation of a service-oriented system is triggered by changes that may occur in three major areas: (a) the expectations that its users (or other stakeholders) have concerning some quality that the system should provide; (b) the world in which the system is executed; (c) the system itself.

The first area is the realm of "non-functional requirements," the second is traditionally called "context," and the third one is often referred to as the "machine" (Jackson 2001).

The reader can notice that another important area where changes occur frequently and that demands changes in software systems is functional requirements. We do not consider it in this chapter as this concern cannot be addressed with self-adaptation, but, instead, requires software systems to undergo a re-development cycle (often called evolution).

In general, all changes that can be addressed by performing some simple reasoning (for example, "since the user is in downtown Milan, he/she is interested in knowing about free parking spaces in that area"), or by modifying the bindings to services, can be addressed by a service-oriented system if it incorporates self-adaptation facilities (de Lemos et al. 2011). Conversely, the changes that require redesign or reimplementation of the system typically have to be addressed by the intervention of human beings and are not considered in this chapter.

The area of non-functional requirements refers to Qualities of Service (QoS) such as the service response time or its availability. During the execution of a service it can happen that the expectation of users for a certain non-functional requirement, e.g. response time, changes. This change may be for various reasons, for instance, to a new regulation that requires users to perform some operation in a shorter time. The change can be, to a certain extent, accommodated through the self-adaptation capabilities of the system, taking into account its execution conditions and its ability to cope with an increase of the response speed for a certain user.

The term context derives from the Latin *cum* (with or together) and *texere* (to weave). It has been defined by Dey and Abowd (2000) as "any information that can be used to characterise the situation of entities (persons, places, objects) that are relevant to the interaction between a user and an application, including the user and the application themselves." According to Hofer et al. (2002), context can be physical, i.e. measured by some hardware sensor, or logical, i.e. captured by monitoring user interaction. When it is physical it refers to location, movement, and any environmental information. When it is logical it refers to users' goals, emotional state, business processes, etc. In both cases, the context evolves during the execution of the service-oriented system. For instance, the user may change location, or the business processes in which he/she is involved can change. Self-adaptation capabilities of the system need, therefore, to be able to sense and to reason about these changes.

Both changes in the non-functional requirements and in context happen in the external environment that, therefore, has to be continuously monitored by the system. Some changes to the system also happen internally, both in its hardware and software resources. An example concerns the increase/decrease of the computational power available for the software system. Other examples of changes that may occur and trigger the need for adapting the service-oriented system are shown in fig. 8.1 above. Clearly, the examples of changes shown in the figure require different levels of intervention on the corresponding software system and its resources.

For instance, the lack of computational resources may be addressed at the infrastructure layer, without modifying the structure of the service-oriented system, for example by exploiting the flexibility offered by cloud computing (Armbrust et al. 2010). The failure of a service is addressed at the service-composition layer through dynamic binding of alternative and compatible services. In this case, in fact, the substitution of one service for another can occur at runtime without performing any reprogramming activity of the software (Moser et al. 2008).

A classification of various kinds of adaptation needs can be found in (Metzger et al. 2011). Among others, the authors identify "reactive adaptation," that is, the ability of the system to react to changes when they occur, and "proactive adaptation," that is, the ability of the system to anticipate the need for adaptation. Connected to this classification, the authors also comment on the importance of managing adaptation across all layers of a software system. We also highlight self-adaptation that copes with problems within the machine can be classified as "self-healing," i.e. the ability to repair the system or "self-optimisation," i.e. the ability to improve the performances of the system.

Fig. 8.1. Areas for changes and examples

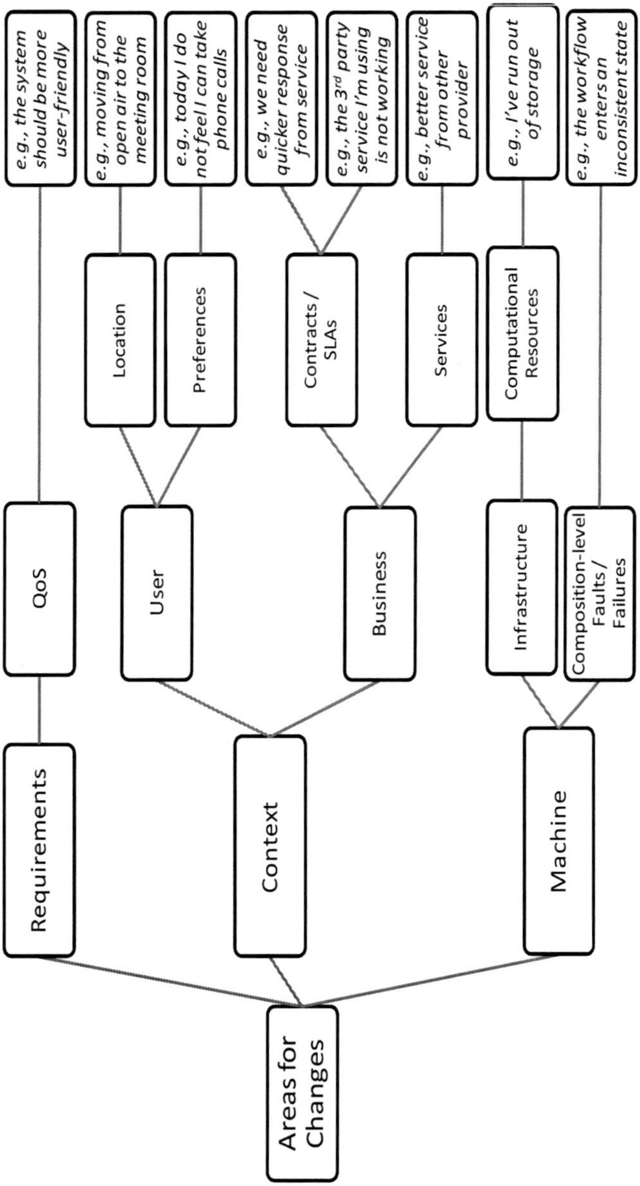

To enable adaptation needs to trigger self-adaptation, the service-oriented system has to be properly built and deployed. At design time the mechanisms that enable self-adaptation have to be prepared. At runtime a monitoring infrastructure has to ensure that the adaptation needs are captured and that proper adaptation strategies are executed. The life cycle presented in the following section focuses on both the design time and the runtime aspects and highlights the main activities and artifacts to be produced in each phase.

3. The S-Cube lifecycle and self-adaptation

As discussed before, at design time, functional and non-functional requirements, context and machine characteristics have to be analysed in order to identify the types of changes that trigger self-adaptation and consequently define the needed mechanisms to monitor the environment and the system behaviour. In this phase, the strategies for self-adaptation have to be also developed. The life cycle proposed in the S-Cube project (see fig. 8.2 below) addresses all these aspects (Bucchiarone et al. 2009) and (Bucchiarone et al. 2010). Such life cycle is composed of two cycles: (a) the "evolution" cycle that leads to the explicit re-design of an application, (b) the "adaptation" cycle that is performed at run time when the application needs to be adapted on-the-fly. The two cycles coexist and support each other during the lifetime of the application.

The initial phase of the life cycle is (Early) Requirement Engineering and Design, in which the adaptation and monitoring requirements are elicited on the basis of the context dimensions and the application and system characteristics that are considered relevant for the adaptation and the evolution of the considered system. Thus, this phase is responsible for the definition of the types of changes that trigger self-adaptation. Successively, during the System Construction phase, the corresponding monitors and the adaptation mechanisms are designed, developed and then refined until the Deployment phase. When the system is running (Operation and Management phase), continuous monitoring is executed supporting the detection of the relevant context and system changes. If changes occur, it is possible to proceed in two different directions: executing evolution or adaptation of the system. Evolution is performed if the system needs to be redesigned and thus it requires the reiteration of the described cycle starting from the requirements engineering and design phase.

Fig. 8.2. Life-cycle for adaptable service-oriented systems

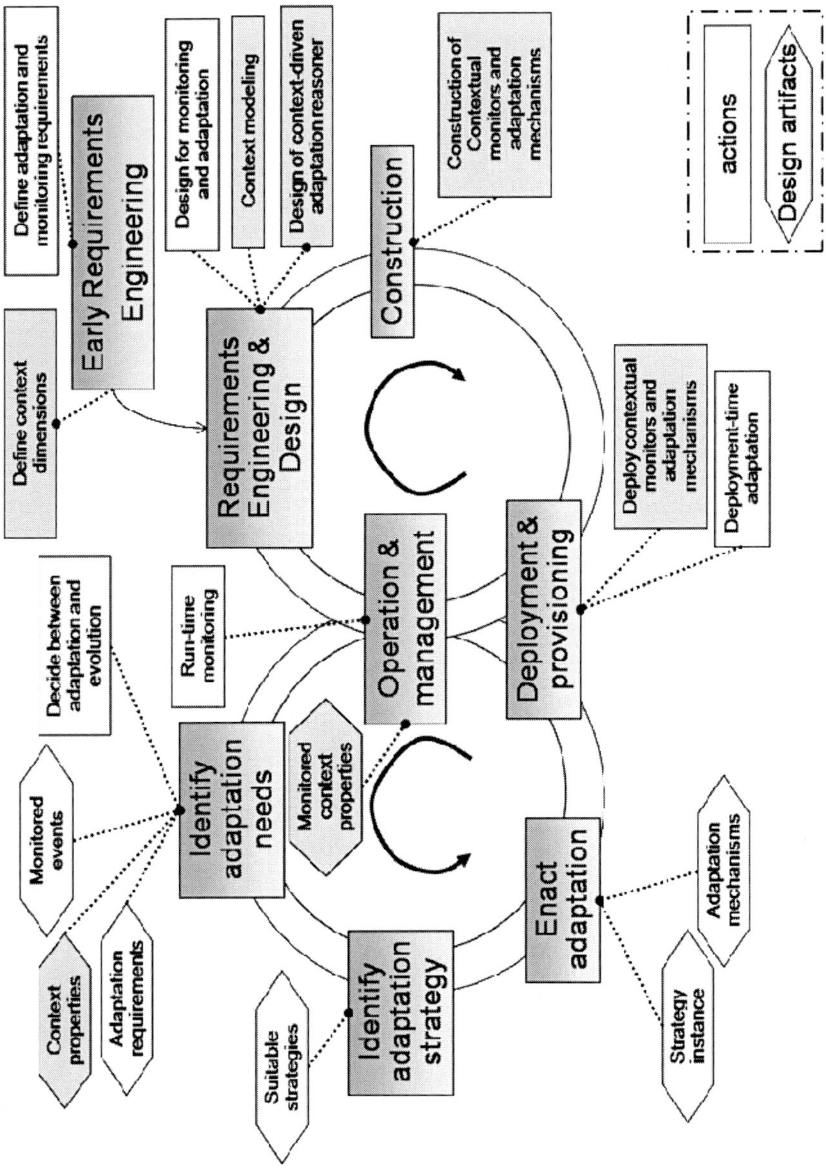

If changes require adaptation, the Adaptation Needs that can be triggered from monitored events and adaptation requirements have to be identified. An adaptation need can be defined as a specific problem-situation that demands for adaptation. Each adaptation need has to be associated with the different Adaptation Strategies able to satisfy the corresponding adaptation requirements. Finally, on the basis of the current situation, the knowledge obtained from previous executions, and the available adaptation mechanisms, a reasoner selects the most suitable adaptation strategy that will be automatically or manually performed in the Enactment phase.

Fig. 8.2 above highlights the various adaptation- and monitoring-specific actions (boxes) carried out throughout the life cycle of a service-oriented system, the main design artifacts that are exploited to perform adaptation (hexagons), and the phases where they are used (dotted lines).

While the life cycle can be applied when reasoning at different layers of a service-oriented system, self-adaptation triggers and strategies relevant for each layer are different. In the following two sections we discuss those that are relevant to the composition layer and the infrastructure layer. Moreover, we discuss the interplay of adaptation actions performed at these two layers.

4. Possible adaptation triggers and actions at the composition layer

In order to offer efficient and reliable service-oriented systems, it is necessary to guarantee that the service components are always aligned with the changing world around them. As we have seen in the previous section, at design time we identify the possible needs that trigger adaptation, and that from now on we call "triggers." Also, we define the possible alternative strategies to support adaptation.

To do this, we need to have a concrete model of the service composition context and of those aspects of the requirements and of the machine that are relevant for triggering the adaptation or evolution of a service composition, and that enable the identification and the collection of the proper context information. As we have seen in fig. 8.2, starting from the requirements analysis phase where, in parallel to the precise definition of requirements, a proper "context model" has to be defined.

Once the context model and its relation with the application model are defined, it is necessary to properly capture and define the adaptation aspects. In particular, it is important to define when the contextual changes are critical for the service composition functioning (i.e. adaptation triggers) and what should be done or achieved when these changes take place (adaptation strategies).

Depending on the context dimension/element and on the specific requirements of a service composition, the definition of the adaptation trigger may vary. As a result, the corresponding context monitors needed to detect those triggers will have different, sometimes application-specific, forms and realisations.

Furthermore, in certain cases the adaptation trigger does not correspond to some value of a particular single context element, but is characterised by the complex combination of different context dimensions/factors. In the simplest case, such situations may be directly encoded using the capabilities provided by existing monitoring frameworks as Dynamo (Baresi et al. 2005). In other cases, a sophisticated reasoner may be necessary. For instance, in Kazhamiakin et al. (2008), a context represents the combination of user's personal assets (agenda, location, social relations, and money), and the critical context changes characterise some critical combination of those assets. A dedicated analysis mechanism is used to identify those situations (asset conflicts) and to trigger adaptation solutions.

As for triggers associated to the component services, considering the context, we can define adaptation actions to align the service composition with its context changes. Considering the contextual aspects, it is necessary to take into account that adaptations may be performed (a) to "customise" the composition in order to fit to the situation in which it currently operates, (b) to "optimise" the composition in order to improve certain (usually QoS) issues and characteristics of the composition, or (c) to "prevent and avoid" future faults or undesirable situations in the execution of the system. The customisation is often driven by changes in the composition context and especially by time, location, user, business, and service context dimensions. Optimisation is more related to the service and computing context dimensions but can be also triggered by changes in the users' preferences. Finally prevention is mostly related to the service and computing context since changes of the execution environment might increase the risk of failures and the need for prevention. Some of these aspects may be also interleaved. For example, if a user moves to a new location (i.e. change in the user context), a new set of services may be available (i.e. change in the business context) with different bandwidth (i.e. change in the machine).

As represented in table 8.1, examples of triggers relevant for the composition layer include changes in the time in which the system operates, in the location of the user, in the user preferences (indicated in the table simply as "User"), in the services being exploited, in the business processes and in the machine. Possible adaptation strategies include the possibility of replacing one or more services with some others, the re-execution of some services, the transformation of the composition itself, for example by introducing new paths of service calls, the complete failure of the composition, the possibility of keeping the service bindings abstract until runtime, the possibility of renegotiating a Service Level Agreement (SLA), the possibility of compensating for failing transactions, and the possibility of leaving the adaptation loop to enter into the evolution loop.

Table 8.1. Suitability of adaptation strategies to react to context changes

Adaptation Strategy	Adaptation Triggers					
	Time	Ambient	User	Service	Machine	Business
Service substitution	X	X	X	X	X	X
Re-execution				X	X	
Re-composition				X	X	X
Fail			X	X	X	X
Service concretization	X	X	X	X	X	X
Re-negotiation			X		X	X
Compensation				X	X	
Trigger Evolution		X	X	X	X	X

As shown in table 8.1, each trigger can be associated with a set of adaptation strategies that are suitable to realign the application within the system and/or context requirements. In order to select the adaptation strategy to apply, it is necessary to consider that adaptation triggers may be associated with other requirements that are important for designing and performing adaptation, in particular: the “scope of the change,” i.e. whether the change affects only a single running instance of the service composition or influences the whole model and, the “impact of the change,” i.e. the possibility of the application to accomplish its current task. Depending on these parameters different strategies may apply. For example, when the scope of the change concerns the whole application model “trigger evolution” strategy applies. As for the impact, such strategies as “re-execution” or “substitution” may apply when the service composition state did not change and the task still can be accomplished.

On the other hand, "compensation," "fail" or "trigger evolution" apply when there is no way to complete the current task.

While the life cycle we have introduced so far can be applicable to both the composition and the infrastructure layer, the kinds of adaptation triggers as well as the corresponding strategies offered by the infrastructure layer are mainly focused on what is happening within the machine.

5. Possible adaptation triggers and actions at the infrastructure layer

In greater detail, adaptation at the infrastructure layer is often triggered by variations in the load of the system (either measured in terms of number of requests per time unit or in terms of CPU cycles needed to complete the jobs in the system queues). In these cases, the infrastructure layer may decide to replicate the system on new machines, for instance, by creating clones of the currently used virtual machines (VMs from now on) or to balance the load among the existing virtual machines.

Another relevant trigger for adaptation concerns the availability of new resources that were not accessible before. In this case, the infrastructure layer may decide to use them to replace the available ones, if they are more convenient, or to use them together with the existing ones.

Some triggers can also be external to the machine. An example is a change in the QoS users require. Such a change can be accommodated directly by the infrastructure layer in case it enables an elastic use of resources.

Table 8.2 shows some possible triggers and strategies. The "X" symbols highlight the suitability of a certain strategy for a specific trigger. In many cases different strategies have to be combined together. For instance, when the load is too high for the system, new machines can be started, but this action alone does not solve the problem. We have also to move or clone some VMs on the newly started machines in order to exploit their capabilities.

Table 8.2. Adaptation triggers and strategies at the infrastructure layer

Adaptation strategy	Adaptation trigger				
	Load is too low / too high	New computing resources are available	One or more computing resources become unavailable	A new cloud becomes more convenient	QoS requirements change
Clone a VM	X				X
Move a VM		X	X	X	
Balance load among VMs	X				
Switch machines on/off	X				X
Move data storage			X	X	
Migrate from a cloud to the other	X	X		X	X

6. Coordinating cross-layer adaptation

As introduced above, service-oriented systems operate on different layers to implement their functionality and to provide their expected quality. We have identified and discussed two main layers: (a) the "infrastructure layer," in which computational and storage resources used for executing the system reside (see the Section on possible adaptation triggers and actions at the infrastructure layer); (b) the "composition layer," in which the services offered by the system are made available and composed (see the section above on possible adaptation triggers and actions at the composition layer). In this two-layer setting, self-adaptation can happen both at the infrastructure and at the service composition layer.

However, if adaptations on those layers are not coordinated, this can lead to the following two important issues: First, conflicting adaptations can arise, for example in cases where the individual adaptations cancel each other or interact in an undesired way. Secondly, there may be cases in which isolated applications on single layers will not be enough to resolve the problem; for example a problem can only be resolved if the composition of the system is changed while at the same time more computing resources are made available in the cloud. We elaborate on both of these issues in the following sub-sections.

6.1 Avoiding Conflicting Adaptations

In Bucchiarone et al. (2010b), a general framework has been proposed to address the harmful side-effects resulting from independently acting adaptations. In Kazhamiakin et al. (2008b), general, adaptation-specific failures and possible quality assurance techniques to uncover those failures have been introduced. Finally, in Guinea et al. (2011), a more concrete framework that integrates layer specific monitoring and adaptation techniques, and enables multi-layered control loops has been proposed for service-based systems. All these contributions lend themselves naturally to discuss the problems that may arise when considering adaptations which are independently performed on different "layers."

The following important classes of problems can arise in the presence of independently acting adaptations:

- **Conflicts:** A conflict arises if each layer reacts to a relevant change by issuing contradictory adaptations concurrently. Contradictory adaptations are adaptations that cannot be applied in combination since they would violate the requirements of the system or would not be able to solve the observed deviation from requirements. In certain situations the timing of events may be the source of another such kind of conflict. For example, it might happen that changes occurring on the infrastructure layer are so fast, that although they can be addressed efficiently on that layer, the impact they have on the composition layer leads to severe problems, as the execution of adaptation activities on the composition layer is not quick enough to follow. This may mean that new events continuously trigger new adaptations, leading to an "adaptation stack overflow," possibly delaying other important adaptations or bringing the application to a halt completely.
- **Oscillations:** An oscillation can occur if conflicting adaptations are issued in sequence. For example, the composition layer may receive an event that triggers an adaptation which in turn leads to a critical change that leads to an adaptation on the infrastructure layer. It might happen that the adaptation of the infrastructure layer

in turn leads to critical event on the composition layer, necessitating yet another adaptation. Obviously, if this undesired behaviour is not observed and controlled, this can lead to a long series of mutual adaptations without ever reaching a stable system configuration, i.e. this can lead to an "adaptation livelock."

- **Race conditions:** Race conditions occur if the two layers apply adaptations and the final outcome depends on the order in which these adaptations are performed and completed. As an example, an adaptation on the composition layer may require the deployment of a new service on the cloud. Yet, if the adaptation of the cloud needed to provide additional resources for that new service is executed after the deployment of the new service, it might happen that the deployment will not succeed due to limited computational resources. Yet, if the resources would have been reserved before deployment, this adaptation would have been successful.

There are basically two complementary ways of avoiding the above problems: (a) one can check whether those problems may actually occur by using targeted quality assurance techniques as proposed by (Cheng et al. 2009; Kazhamiakin et al. 2008b); (b) one can restrict the freedom of each individual adaptation by coordinating their adaptation actions as proposed by (Guinea et al. 2011; Zengin et al. 2011; Bucchiarone et al. 2010b; Popescu et al. 2010; Cheng et al. 2009; Vidackovic et al. 2009).

In the remainder of this chapter, we will follow the latter approach and will provide a concrete example for how such a coordination of adaptation actions can be realised.

6.2 Exploiting Synergy of Adaptations

Of course, the problems that can arise due to uncoordinated adaptations are numerous. Thus, one might be tempted to restrict adaptation to one single layer only in order to avoid these problems. However, there may be situations in which only the execution of concurrent adaptations on more than one layer can address the problems faced.

In an example cited by Schmieders et al. (2011), the coordination of adaptations on the infrastructure and the composition layer can increase the chance that performance violations of the application can be avoided. For instance, a slow service on the composition layer may be replaced by a faster alternative service. However, this might not be enough to achieve the expected response time of the application. Thus, an adaptation on the infrastructure layer may additionally be used to reserve more computing resources, thereby increasing the responsiveness of services. As this example shows, there may be situations in which a coordinated combination of adaptations can lead to more situations in which requirements violations can be avoided.

This chapter will not further scrutinise this specific aspect of cross-layer adaptation. However, we acknowledge the fact that understanding how to perform such kinds of adaptation should certainly be part of future research.

7. Use Case

In this section, we describe how self-adaptation can be implemented in the life-cycle of challenging Real-Time Online Interactive Applications (ROIA) in cloud environments. Examples of ROIA are multi-player online computer games, interactive e-learning and training applications and high-performance simulations in virtual environments. We illustrate the specific features of ROIA and express their major design and execution aspects in the context of the S-Cube Lifecycle model. Our use case focuses on the execution of ROIA on computational resources and, thus, illustrates adaptation scenarios applied mostly at the infrastructure layer, as described in the previous sections.

In ROIA, there are typically multiple users who access a common application state and interact with each other concurrently within one virtual environment. The users connect to the application from different client machines and interact with other users: for example, in online games users move their avatars through a virtual environment and interact with other users' avatars or computer-controlled characters ("entities"). Since ROIA usually have very high performance requirements, the application processing is performed on multiple servers. Hence, ROIA are highly distributed applications with challenging QoS demands, such as: short response time to user actions (about 0.1-1.5 seconds), high update rate of the application (up to 50 Hz), and large and frequently changing number of users in a single application instance (up to 10^4 simultaneously). To address these demands, application developers need to implement suitable adaptation mechanisms. In our previous work we studies how the Lifecycle model and self-adaptation are applied to the challenging domain of ROIA Meiländer et al. (2010).

A major problem for an efficient ROIA execution is the economical utilisation of server resources, which is difficult due to a variable and continuously changing number of users. This leads to expensive up-front investments to build a suitable server pool which is able to handle peak user numbers but will be underutilised most of the time when the load is below the peak.

Cloud computing offers new opportunities for ROIA execution and promises a potentially unlimited scalability by distributing application processing on an arbitrary number of resources given suitable adaptation mechanisms. Clouds allow for adding/removing resources on demand. This opens for ROIA an opportunity to serve very high numbers of users and still comply with QoS demands. Despite a variable number of users, cloud resources can be used efficiently if the application supports adding/removing resources during run-time. Hence, using cloud computing for resource provision and the Lifecycle model for implementing adaptable ROIA complement each other.

In order to support ROIA development and adaptation on clouds, we have developed the RTF-RMS resource management system (Meiländer et al. 2011) on top of the Real-Time Framework (RTF) (2011). RTF-RMS implements the following mechanisms for ROIA development on Clouds:

- **Monitoring** of application-specific data, e.g. update rate, number of entities, etc.
- **Distribution handling** for the dynamic adaptation of running application sessions by adding/removing cloud resources on demand using particular adaptation strategies (described below).
- **Application profiles** that allow developers to specify application-specific adaptation triggers.
- High-level **development support** for communication handling and application state distribution.

In fig. 8.3 below, we illustrate how RTF-RMS supports the developer in designing adaptable ROIA according to the different phases of the Lifecycle model; the phases are numbered in the figure.

In the Requirement Engineering phase of the Lifecycle model, the application developer must identify suitable adaptation requirements for his application. In the case of ROIA, the mechanisms for monitoring and adaptation should be non-intrusive, i.e. take place in parallel with the application execution, such that users are not aware of changes inside the application.

Fig. 8.3. Mechanisms of RTF-RMS for ROIA development according to the Lifecycle model

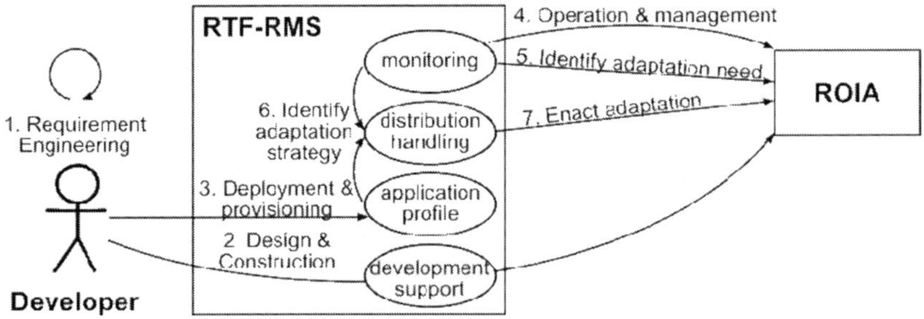

For the Construction phase, RTF-RMS provides the developer with a C++ library of high-level functions for optimised communication handling (client-server and inter-server) and efficient application state distribution in a multi-server environment. By using RTF-RMS for communication and distribution handling, monitoring is automatically integrated in the application processing. Monitoring data is used in the next phase of the Lifecycle to implement adaptation triggers, as described in the previous sections. For ROIA, we distinguish between the following adaptation triggers:

- Change in QoS, for example caused by unreliable hoster resources
- Change in the machine, for example caused by increasing user interactions, making computation of state updates more expensive
- Change in the business context, for example more users connect to the application due to changing user preferences.

In the Deployment and Provisioning phase, trigger rules are defined in the RTF-RMS application profile for each adaptation trigger, (see fig. 8.4 for an example). In the application profile, the developer specifies thresholds for the monitoring data provided by RTF-RMS. Adaptation is triggered if corresponding monitoring values pass these thresholds. In the example, adaptation is triggered if the update rate drops below 25 Hz or exceeds 50 Hz. The application developer can find suitable values for all parameters in the application profile by considering the runtime behaviour of the application on physical resources and calculating the virtualisation overhead, or by conducting benchmark experiments in a Cloud environment.
Fig. 8.4. Example application profile for a fast-paced action game.

```
<appProfileData>
 <metric>
  <name>UpdateRate</name>
  <addResourceThreshold>25</addResourceThreshold>
  <removeResourceThreshold>50</removeResourceThreshold>
 </metric>
</appProfileData>
```

In the Operation and Management phase, the application is running and monitoring data is checked continuously against the trigger rules to detect changes in the context or in the system that could require adaptation.

In the Identify Adaptation Need phase, RTF-RMS detects a violation of trigger rules which indicates, for example, an overload of the application session: the update rate drops below 25 Hz.

In the Identify Adaptation Strategy phase, RTF-RMS analyses the number of application servers and their current workload to choose an adaptation strategy (Meiländer et al. 2011; the cited paper refers to adaptation strategies as load-balancing actions). The adaptation strategies are based on the partitioning of the virtual environment among servers. In particular, the processing of entities inside a particular area (zone) is assigned to different servers. RTF-RMS implements currently the following three adaptation strategies:

- **User migration**: Users are migrated from the overloaded server to an underutilised server processing the same zone. For this purpose, user connections are switched from one server to another. RTF-RMS distributes users by default equally between the application server for a particular zone. User migration is the preferred action if the load of an overloaded server can be compensated by currently running resources.
- **Replication enactment**: New application servers are added in order to provide more computation power to the highly frequented zone. This strategy is called replication: each application server keeps a complete copy of the application state, but each server is responsible for computing a disjoint subset of entities. After enacting replication, RTF-RMS migrates a number of users to the new replica in order to balance the load. Replication implies an additional inter-server communication, so its scalability is limited. If the number of active replicas for a particular zone is below the maximum number of replicas specified by the application profile,

replication is used to add new resources; otherwise the resource substitution strategy (described next) is preferred.

- **Resource substitution**: an existing resource in the application processing is substituted by a more powerful one in order to increase the computation power for highly frequented zones. For this purpose, RTF-RMS replicates the targeted zone on the new resource and migrates all clients from the substituted server to the new server. The substituted server is then shut down.

In the Enact Adaptation phase, RTF-RMS enacts the chosen adaptation strategy and changes the distribution of the application processing accordingly.

An important factor that may limit the performance of applications on clouds is the long startup time of cloud resources which may take up to several minutes. Since ROIA are highly responsive applications, finding a compensation for long startup times is an important task. For this purpose, RTF-RMS implements a resource buffer to which a predefined number of cloud resources are moved in advance, i.e. before they are demanded by the application. The number of buffered resources is configured in the application profile and has to be chosen carefully since resource buffering generates additional costs. A more detailed discussion on how to choose the size of the resource buffer can be found in Meiländer et al. (2011).

Another critical issue for the cost-efficient ROIA provision on clouds is the consideration of leasing periods. In commercial cloud systems, resources are typically leased and paid per hour or some longer leasing period. Since ROIA have dynamically changing user numbers, the time after which cloud resources become dispensable is very variable. However, resources will not be used cost-efficiently if they are shut down before the end of their leasing period. Hence, RTF-RMS removes the resources that have become dispensable from the application processing and moves them to the resource buffer. Cloud resources in the buffer are shut down at the end of their leasing period or they are integrated in the application processing again if required.

In order to demonstrate the influence of adaptation using RTF-RMS on the application performance, we present experimental results of the user migration adaptation strategy using an example of a multi-player action game called RTFDemo (RTF 2011). In order to provide a seamless gaming experience, users should not receive less than 25 updates per second over a longer time period. Hence, we defined an adaptation trigger rule with 25 updates per second as the lower threshold.

Fig.8.5. Adaptation using user migration

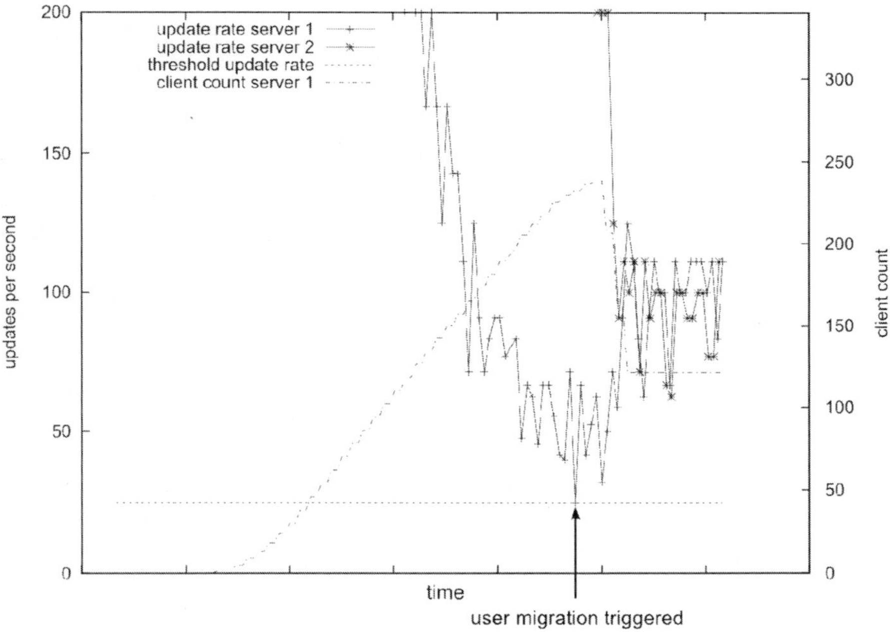

For our experiments, we use a private cloud environment with the Eucalyptus framework (version 2.0.2) (Nurmi et al. 2009). We have started two application servers on cloud resources for replication of a single zone. All clients are initially connected to server 1, which implies an imbalanced load on application servers. We apply adaptation at the infrastructure level by balancing load among virtual machines using user migration. The load is balanced by assigning the processing of half of the users that are connected to server 1 to server 2. The corresponding user connections are switched between these Cloud resources accordingly. Fig. 8.5 above shows that the update rate of server 1 initially drops with the growing number of connected clients. When the update rate reaches the threshold of 25 Hz for 240 users, RTF-RMS migrates half of the users (i.e. 120 users) from server 1 to server 2. This adaptation strategy increases the update rate of server 1 from 25 Hz to about 100 Hz.

8. Related work

The literature contains several approaches for the definition of a life cycle for the design and management of service-based applications. Earlier approaches such as ASTRO (Trainotti et al. 2005), focused on the possibility of monitoring and intervening on SBAs in order to recover from unwanted and unexpected behaviour and assume human interventions. More recent work tried to handle this issue proposing methods for developing self-adaptive applications. As suggested in (Ardagna and Pernici 2007), adaptation mechanisms can either be embedded in the description of the adaptable SBAs or implicit in its structure. Various frameworks can be found in the literature with the objective of supporting adaptation of SBAs. In Linner et al. (2008), authors propose a life-cycle supporting self-adaptation of the service-based applications even if they lack of explicit guidelines for their design. In any case, most of the literature contributions concern built-in adaptation, i.e. the adaptation logic is completely specified at design time. They focus on the specification of adaptation mechanisms and adaptable applications, exploiting different tools. For instance, the specification may be performed by extending standard notations (i.e. BPEL) with adaptation specific tools (Karastoyanova et al. 2005), event conditions actions rules (Baresi et al. 2007), variability modeling (Hallerbach et al. 2008) or aspect-oriented approaches (Kongdenfha et al. 2006).

In the literature there are, however, proposals of frameworks for dynamic adaptation, all featuring an adaptation manager separated from the application. Notably, all these approaches are in the service-oriented field. In Spanoudakis et al. (2005) the authors consider the problem of adapting the application by replacing malfunctioning services at runtime. In this approach, adaptation strategies are triggered as a consequence of a requirement violation. In fact, the adaptation rule is fixed at design time, but is dynamically applied by a manager component that monitors functional and non-functional properties, creates queries for discovering malfunctioning services and replaces them with dynamically discovered replacements. In other approaches, adaptation is triggered by application constraints. For example, Narendra et al. (2007) propose an aspect-oriented approach for runtime optimisation of non-functional QoS measures. QoS constraints are also the basis for dynamic reconfiguration of processes in the METEOR-S framework (Verma et al. 2005). Reconfiguration is performed essentially at deployment-time. Some more general frameworks are also available. An example is SCENE (Colombo et al. 2006) where event-condition-action rules are used for defining different kinds of self-adaptation rules dealing with runtime replacement of services, service re-execution, renegotiation of SLAs, replacement of parts of a service composition, etc. As in other cases, in this approach rules are triggered during the execution as a result of a monitoring activity and they can trigger various dynamic actions that change the service composition depending on many variables, including the status of the context. While a set of basic rules is defined at design time, new rules can also be added during the execution of the system as a result of a learning activity.

PAWS (Ardagna et al. 2007) is a framework for flexible and adaptive execution of web service based applications. At design-time, flexibility is achieved through a number of mechanisms, i.e. identifying a set of candidate services for each process task, negotiating QoS, specifying quality constraints, and identifying mapping rules for invoking services with different interfaces. The runtime engine exploits the design-time mechanisms to support adaptation during process execution, in terms of selecting the best set of services to execute the process, reacting to a service failure, or preserving the execution when a context change occurs. Finally the Vienna Runtime Environment for Service-oriented Computing (VRESCo) (Hummer et al. 2011) is a framework implemented to address issues like dynamic selection, binding and invocation of services.

In general, adaptable systems change their behaviour, reconfigure their structure and evolve over time reacting to changes in the operating conditions, so as to always meet users' expectations.

Nowadays, recent literature contributions started to address adaptation issues in the contexts of grid or cloud computing. All existing contributions focus on specific aspects. In Brandic et al. (2009), authors considers cloud computing stating that such environments require adaptable services that can cope with system failures and environmental change minimising human intervention. In particular, this chapter mainly proposes models for service negotiation and SLA mapping. Lim et al. (2009) instead considers "elastic cloud computing APIs" as the natural opportunity for designing controllers able to automate an adaptive service and resource provisioning, and many recent works have explored feedback control policies for a variety of network services under various assumptions. The approach presented in this chapter aims at providing a wider support in the design and adaptation of service based applications executed in cloud computing contexts.

9. Conclusion

In this chapter we have discussed the needs that trigger self-adaptation in service-oriented systems and we have seen that adaptation can occur at different layers of the system architecture and, in some cases, that it can also encompass different layers. The life cycle offered by the S-Cube project is a helpful tool in guiding developers in the identification of those self-adaptation triggers and strategies that are relevant for the specific system under development, and allowing providers to offer the proper runtime monitoring and reasoning mechanisms needed to execute the defined adaptation strategies.

Acknowledgements

Research leading to these results has received funding from the European Community's 7th Framework Programme FP7/2007-2013 under grant agreement 215483 (S-Cube).

References

Ardagna, D. and Pernici, B. "Adaptive Service Composition in Flexible Processes." *IEEE Transactions on Software Engineering* 33 (6) (2007): 369–384.

Ardagna, D., Comuzzi, M., Mussi, E., Pernici, B. and Plebani, P. "PAWS: a framework for processes with adaptive web services." *IEEE Software* 24 (6) (2007): 39–46.

Armbrust, M., Fox A., Griffith R., Joseph A. D., Katz R., Konwinski A., Lee G., Patterson D., Rabkin A., Stoica I. and Zaharia M. "A view of cloud computing." *Communication ACM* 53 (2010): 50–58.

Baresi, Luciano and Guinea Sam. "Dynamo: Dynamic Monitoring of WS-BPEL Processes." ICSOC, 2005.

Baresi, L., Guinea, S. and Pasquale, L. "Self-healing BPEL processes with Dynamo and the JBoss rule engine." In Proceedings of ESSPE 2007, 11–20. New York: ACM Press, 2007.

Brandic, I., Music, D. and Dustdar, S. "Service mediation and negotiation bootstrapping as first achievements towards self-adaptable grid and cloud services." In Proceedings of the 6th international conference industry session on Grids meets autonomic computing (GMAC '09), 2009.

Bucchiarone, A., Cappiello, C., Di Nitto, E., Kazhamiakin, R., Mazza, V. and Pistore, M. "Design for Adaptation of Service-Based Applications:

Main Issues and Requirements." Proceedings of ICSOC, ServiceWave Workshops, 2009.

Bucchiarone, A., Kazhamiakin, R., Cappiello, C., Di Nitto, E. and Mazza V. "A context-driven adaptation process for service-based applications." In Proceddings of PESOS, 2010.

Bucchiarone, A., Marconi, A., Pistore, M., Föll, S., Herrmann, K., Hiesinger, C. and Marinovic, S. "An Overall Process for Self-Adaptive Pervasive Systems." In Proceedings of the Second International Conference on Adaptive and Self-adaptive Systems and Applications (ADAPTIVE 2010), Elsevier, 2010b.

Cheng, B. et al. "Software engineering for self-adaptive systems: A research roadmap." In *Software Engineering for Self-Adaptive Systems*, edited by B. Cheng, R. de Lemos, H. Giese, P. Inverardi, and J. Magee. New York: Springer, 2009.

Colombo, M., Di Nitto, E. and Mauri, M. "SCENE: A Service Composition Execution Environment Supporting Dynamic Changes Disciplined Through Rules." In ICSOC 2006, edited by Dan, A., Lamersdorf, W., LNCS, vol. 4294, 191–202. New York: Springer, Heidelberg, 2006.

De Lemos, R. et al. "Software Engineering for Self-Adpaptive Systems: A second Research Roadmap." In Software Engineering for Self-Adaptive Systems, number 10431 in Dagstuhl Seminar Proceedings, Dagstuhl, Schloss Dagstuhl, edited by R. de Lemos, H. Giese, H. Müller and M. Shaw, Leibniz-Zentrum fuer Informatik, Germany.

Dey, A. K. and Abowd, G. D. "Towards a Better Understanding of Context and Context-Awareness." In Workshop on The What, Who, Where, When, and How of Context-Awareness, 2000.

Di Nitto, E., Ghezzi, C., Metzger, A., Papazoglou, M. and Pohl, V. A "Journey to Highly Dynamic, Self-adaptive Service-based Applications." *Automated Software Engineering.* 2008.

Guinea, S., Kecskemeti, G., Marconi, A. and Wetzstein, B. "Multi-layered Monitoring and Adaptation." In ICSOC 2011, LNCS. New York: Springer, 2011.

Hallerbach, A., Bauer, T. and Reichert, M. "Managing Process Variants in the Process Life Cycle." In Proceedings of ICEIS, 3-2, 154–161. 2008.

Hofer, T., Schwinger, W., Pichler, M., Leonhartsberger, G. and Altmann, J. "Context-awareness on mobile devices: the Hydrogen approach." In 36th Annual Hawaii International Conference on System Sciences, 292–302, 2002.

Hummer, W., Leitner, Ph., Michlmayr, A., Rosenberg, F. and Dustdar, S. "VRESCo—Vienna Runtime Environment for Service-oriented

Computing." In *Service Engineering: European Research Results*, edited by S. Dustdar and F. Li, 299–324. New York: Springer, 2011.

Jackson, M. *Problem Frames: Analysing and Structuring Software Development Problems.* New York: Addison-Wesley, 2001.

Josuttis, N. *SOA in Practice: The Art of Distributed System Design.* Sebastopol: O'Reilly Media, 2007.

Karastoyanova, D., Houspanossian, A., Cilia, M., Leymann, F. and Buchmann, A.P. "Extending BPEL for run time adaptability." In Proceedings of EDOC 2005, 15–26. IEEE Press, Los Alamitos, 2005.

Kazhamiakin, R., Bertoli, P., Paolucci, M., Pistore, M., and Wagner, M. "Having Services 'YourWay!': Towards User-Centric Composition of Mobile Services." FIS, 2008.

Kazhamiakin, R., Metzger, A. and Pistore, M. "Towards correctness assurance in adaptive service-based applications." In ServiceWave 2008, ser. LNCS, no. 5377. New York: Springer, 2008b.

Kongdenfha, W., Saint-Paul, R., Benatallah, B. and Casati, F. "An Aspect-oriented Framework for Service Adaptation." In ICSOC 2006. LNCS, vol. 4294, edited by Dan, A. and Lamersdorf, W., 15–26. New York: Springer, Heidelberg, 2006.

Lim, H. C., Babu, S., Chase, J. S. and Parekh, S. S. "Automated Control in Cloud Computing: Challenges and Opportunities." In Proceedings of the 1st workshop on Automated control for datacenters and clouds (ACDC '09). ACM, New York, NY, USA, 2009.

Linner, D., Pfeffer, H. Radusch, I. and Steglich, S. "Biology as Inspiration Towards a Novel Service Life-Cycle." In Autonomic and trusted computing: 4th international conference. ATC, 2007.

Meiländer, D., Gorlatch, S., Cappiello, C., Mazza, V., Kazhamiakin, R. and Bucchiarone, A. "Using a Lifecycle Model for Developing and Executing Adaptable Interactive Distributed Applications." In *Towards a Service-Based Internet*, 175–186. New York: Springer, 2010.

Meiländer, D., Ploss, A., Glinka, F. and Gorlatch, S. *A Dynamic Resource Management System for Real-Time Online Applications on Clouds.* New York: Springer, 2011.

Metzger, A. and Cassales Marquezan, C., "Future Internet Apps: The next wave of adaptive service-oriented systems?" In ServiceWave 2011, ser. LNCS. New York: Springer, 2011.

Moser, O., Rosenberg, F. and Dustdar, S. "Non-intrusive monitoring and service adaptation for ws-bpel." In Proceeding of the 17th international conference on World Wide Web, WWW '08, 815–824. New York, ACM. 2008.

Narendra, N., Ponnalagu, K. Krishnamurthy, J. and Ramkumar, R. "Run-Time Adaptation of Non-functional Properties of Composite Web Services Using Aspect-Oriented Programming." In *Proceedings of International Conference on Service-Oriented Computing, ICSOC*, edited by Krämer, B., Lin, K. and Narasimhan, P., 546 – 557. New York: Springer / Berlin: Heidelberg, 2007.

Nurmi, D., Wolski, R., Grzegorczyk, C., Obertelli, G., Soman, S., Youseff, L. and Zagorodnov, D. "The Eucalyptus Open-Source Cloud-Computing System." In 9th IEEE/ACM International Symposium on Cluster Computing and the Grid, IEEE Computer Society, 2009.

Papazoglou, M., Pohl, K., Parkin, M. and Metzger, A., (eds). *Service Research Challenges and Solutions for the Future Internet: S-Cube—Towards Mechanisms and Methods for Engineering, Managing, and Adapting Service-Based Systems*. New York: Springer, 2010.

Popescu, R., Staikopoulos, A., Liu, P., Brogi, A., and Clarke, S. "Taxonomy-driven adaptation of multi-layer applications using templates." In Self-Adaptive and Self-Organizing Systems (SASO), 2010 4th IEEE Internatinal Conference, 2010.

Real-Time-Framework. 2011, http://www.real-time-framework.com (accessed October 14, 2011).

Salehie, M. and Tahvildari, L. "Self-Adaptive Software: Landscape and Research Challenges." *ACM Transactions on Autonomous and Adaptive Systems* 4 (2) (2009).

Schmieders, E., Micsik, A., Oriol, M., Mahbub, K. and Kazhamiakin, R. "Combining SLA Prediction and Cross Layer Adaptation for Preventing SLA Violations." 2nd WOSS Workshop. 2011, to appear in Scalable Computing: Practice and Experience, December 2011.

Spanoudakis, G., Zisman, A. and Kozlenkov, A. "A Service Discovery Framework for Service Centric Systems." Services Computing, IEEE International Conference 2005.

Trainotti, M., Pistore, M., Calabrese, G., Zacco, G., Lucchese, G., Barbon, F., Bertoli, P. and Traverso, P. "ASTRO: Supporting Composition and Execution of Web Services." In ICSOC 2005.

Verma, K., Gomadam, K., Sheth, A. P., Miller, J. A. and Wu, Z. "The METEOR-S Approach for Configuring and Executing Dynamic Web Processes." Technical report, University of Georgia, Athens, 2005.

Vidackovic, K., Weiner, N., Kett, H. and Renner, T. "Towards business-oriented monitoring and adaptation of distributed service-based applications from a process owner's viewpoint." In ICSOC/ ServiceWave Workshops. New York: Springer, 2009.

Zengin, A., Marconi, A. and Pistore, M. "CLAM: Cross-layer Adaptation Manager for Service-Based Applications". In: QASBA Workshop @ ECOWS 2011, Lugano, CH, 2011.

Zheng, Z., Zhang, Y. and Lyu, M.R. "Distributed QoS evaluation for real-world web services." In Proceedings of the 2010 IEEE International Conference on Web Services. ICWS '10, IEEE Computer Society, Washington, 2010.

CHAPTER NINE

SERVICE LEVEL MANAGEMENT AND SERVICE DELIVERY MANAGEMENT: CHALLENGES IN FEDERATED E-INFRASTRUCTURES

MATTI HEIKKURINEN AND OWEN APPLETON

Providing an IT service involves more than selecting and deploying best technologies. Independent of the level of their complexity, IT services include—in addition to the technical aspects of the service—legal, administrative and managerial aspects. Outside the small niches of research test-beds or "beta" services, these non-technical aspects are as important as the technical ones. Without addressing them, the technology cannot be used to provide a service to real customers—customers who themselves have to comply with different legal, contractual and administrative regulations.

These non-technical issues become especially challenging when applied in environments where the services are composed of component "sub-services" that may belong to different administrative domains. In these situations, it is difficult to determine what the performance and reliability characteristics of the overall service are going to be. Due to these difficulties in guaranteeing performance, availability and security aspects of such amalgams, most of the services of this kind have up to now used informal service level agreements (the best effort being "service offered as-is").

By limiting themselves to this primitive service level management, these federated services effectively confine themselves to application domains where the explicit or implicit *caveat emptor* can be assumed to be sufficient protection against complaints (and potential legal action) for the service providers. In academia this may well be sufficient initially, but as the services mature such uncertain promises of service quality become limiting.

Both grids and multi-cloud solutions are being studied and applied in areas where the added value provided by the service, or the cost of failure (financial or even human), is high. This means that there is more and more demand for someone to assume the responsibility for the quality of service issues. In addition to above mission critical applications, these infrastructures are also used more and more in an indirect manner, through a familiar user interface or an API. This means that in many cases the end users are not aware of the changes in the underlying platform, and may have different *de facto* expectations of the availability of the service from the ones described in the fine print of the Terms and Conditions. Thus, in order to move these federated computing platforms from technical studies and niche solutions to widely used services, they need to be able to provide similar levels of Quality-of-Service (QoS) assurance as the dedicated solutions they are intending to replace.

To reach the goal of matching the QoS of dedicated solutions, it is necessary to develop a cross-organisational model for the "everything in addition to technology" that this chapter started with. This model should not be dependent on particular features of the technologies, rather it would need to deal with managerial practices that are used to design and complement the automated monitoring and recovery mechanisms, and to make the responsibilities and liabilities clear for everyone involved.

One of the projects aiming at contributing to this vision is gSLM: Service Delivery and Service Level Management in grid Infrastructures (http://gslm.eu), which is trying to merge the practical experience from the provision of distributed services in the grid infrastructures with the traditional IT service management. The results are also studied in the context of the emerging multi-cloud solutions.

1. Introducing IT service management

IT service management is a discipline concerned with the management of IT systems, services and products. It is process-centric and customer focused, in contrast to technologically focused approaches to managing information technology. It does not pertain to any single technology or implementation, rather it provides a framework that IT services, staff processes and behaviours, and customer interactions can be hung on.

It is very widely used, not only in the core IT sector, but also by most major organisations that deal with IT systems as part of their operations. Effectively, almost any large-scale commercial organisation will have a formalised approach to IT service management.

1.1 ITIL

ITSM is dominated by a number of frameworks and best practices that provide guidance on all aspects of service planning, provision and management. Foremost among these is ITIL (http://www.itil-official site.com/) short for the Information technology Infrastructure Library (though ITIL is increasingly treated as a name or brand rather than an acronym). ITIL was developed by the UK Government Central Computer and Telecommunications Agency, since absorbed by the Office of Government Commerce (OGC), which since June 2010 forms part of the Efficiency and Reform group of the Cabinet Office (2011). The common aspect of all of these organisational hosts is that they are (or have been) tasked with supporting public procurement acquisition in order to maximise the benefit of government and public sector spending.

In the 1980s the British government recognised that a growing dependence on IT combined with a lack of standards was leading to a plethora of IT management processes and practices being implemented across the UK public sector. It sought to develop a set of recommendations aimed at harmonising IT management. Based on these goals and background, it is understandable that the focus of the ITIL is not solely on maximising the technical performance of the system (based on some benchmark suite). Instead, ITIL provides tools for making the management of the whole IT system lifecycle more predictable and efficient.

ITIL is released as a series of books covering various aspects of IT management, built around a single view of management based on a process-model. In early versions, the number of volumes reached around thirty, but subsequent versions of ITIL have tried to group or reduce this to a more manageable number. ITIL V3, the latest increment, is structured around twenty-six key processes explained in five volumes.

ITIL v3's key processes are summarised in fig. 9.1 below. The system is iterative, with a cycle between the three main components that supports continual process improvement. SLM in ITIL flows from the premise that Service Level Agreements or other agreements that capture real requirements lead to a service that will be aligned with customer business requirements and therefore meet their needs and requirements. However, SLM does not exist in isolation. It is dependent on other processes, such as service catalogues and portfolios, which define the services that SLM seeks to agree and manage.

Fig. 9.1. ITIL v3 core components

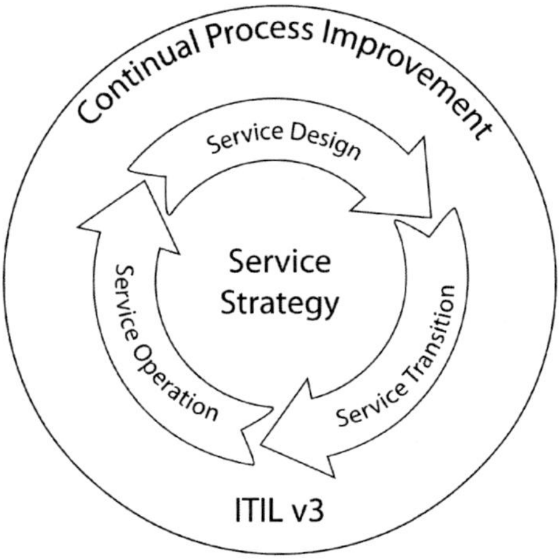

ITIL describes SLM not as an aspirational but as an achievable process. It involves factors that can be monitored and measured, and though it involves softer factors, it uses hard factors for defining agreements. One reason it is able to do this is that it assumes a relatively clear chain of responsibilities, such as that seen in fig. 9.1 above. As we shall see, in the e-Infrastructure domain, the more complex web of relationships is a major complicating factor in adopting service management.

1.2 ISO/IEC 200000

The other major standard in the field is ISO/IEC 20000 (ISO 2011). It was developed by the International Standardisation organisation (ISO). Released in 2005 and updated in 2011, it updates and supersedes BS 15000, previously created by the British Standards Institute. ISO/IEC 20000 was based on the guidance provided by ITIL (and could be said to be compatible with ITIL), but is also compatible with other frameworks.

ISO/IEC 20000 initially presents a Service Management System (SMS). The standard defines the role of the SMS as:

> It specifies requirements for the service provider to plan, establish, implement, operate, monitor, review, maintain and improve an SMS. The requirements include the design, transition, delivery and improvement of services to fulfil service requirements.

Fig. 9.2. PCDA methodology (Diagram by Karn G. Bulsuk, http://blog.bulsuk.com, Licenced as Creative Commons Attribution 3.0 Unported)

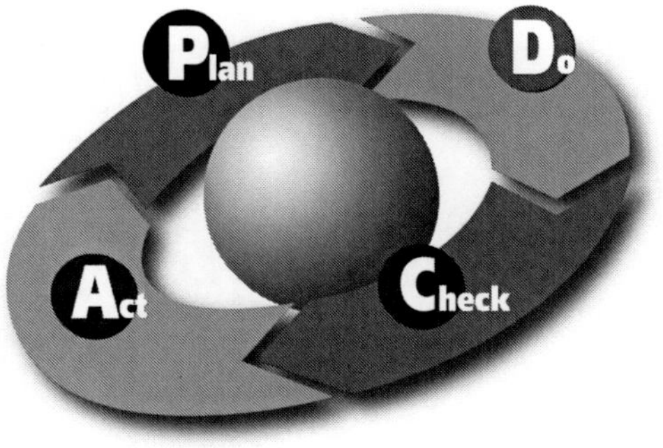

ISO/IEC20000 uses a PDCA methodology—Plan, Do, Check, Act (see fig. 9.2 above).

This is an iterative process where new solutions or approaches are set out, implemented, monitored and tuned. While this may seem a rather generic approach, in some ways it typifies ISO-IEC 20000, which describes itself as "intentionally independent of specific guidance. The service provider can use a combination of generally accepted guidance and its own experience" (ISO 2011).

1.3 Service Level and Service Delivery Management

The overall scope of the ITSM processes can be divided to two categories: Service Level Management (SLM) and Service Delivery Management (SDM). SLM can be seen as linking the ITSM processes to the overall strategic processes of the organisation, whereas SDM focuses on steering and measuring operational level aspects (as name implies).

Within ITIL, SLM is one of several components of Service Design. It:

> … negotiates, agrees and documents appropriate IT service targets with the business, and then monitors and produces reports on delivery against the agreed level of service.
>
> The purpose of the SLM process is to ensure that all operational services and their performance are measured in a consistent, professional manner throughout the IT organization, and that the services and the reports produced meet the needs of the business and customers."
>
> (Cartlidge and Lillycrop 2007)

SLM generally results in a set of agreements that capture the service agreement in a mutually agreed and enforceable format. These are shown in fig. 9.3.

Fig. 9.3. ITSM contract classes

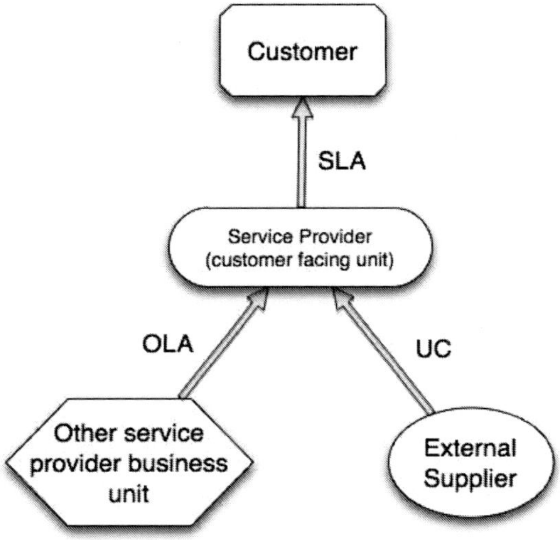

Of these agreements, the most well known (and arguable most important) is the Service Level Agreement. Essentially, Operational Level Agreements (OLAs) are internal versions of SLA within an organisation, while Underpinning Contracts (UCs) are OLAs with external groups that provide components of complete services. While SLM goes considerably beyond just the creation of these agreements, their creation remains a key component.

Service Delivery Management, in contrast, concerns itself with ensuring the service provided matches the agreements made through service level management. This relies on the capability to measure performance of a service provider and compare the results to the agreements around it.

One crucial aspect of both areas, and indeed of much of ITSM in general, is that it does not simply seek to describe services in terms of finances. Rather they look at the creation and provision of value to a customer. While potentially rather nebulous, the concept of value is transferable in a way a currency is not. As will be seen later, it also offers a better chance of describing e-Infrastructure service management than a purely financial basis. Furthermore, the concept of value allows conversion into other “currencies” if needed, for instance in pricing services or scaling penalties for service failure.

In addition to enabling efficient ITSM, a consistent conceptual framework around costs, pricing and value is important also from the contractual standpoint. In many countries, agreements are only enforceable if they represent a “meeting of the minds” between the service provider and customer. In brief, if it is seen that as a provider, you have tried to fool a customer by describing a service in a way that has led them to believe it is not what it really is, any contract around this misrepresented service may not be enforceable. So while the service level agreements negotiated between parties have legal elements and can specify technical metrics, the overall SLM and SDM approaches pay attention to reaching common understandings and other human elements in the service domain. This way they can also be seen as creating a foundation for aligning the IT management with the business objectives or the organisations involved (or supporting regulatory compliance or execution of the core mission/vision in case of non-profit organisations).

Fig. 9.4. The BDIM process closed loop, after (Bartolini et al. 2011)

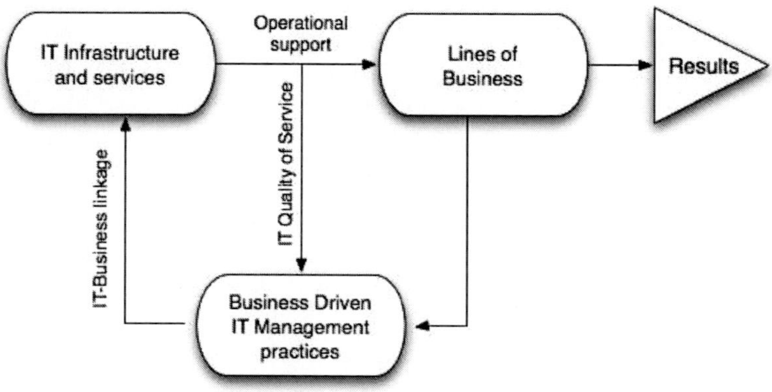

Within the ITSM sector there is a relatively recent approach to ITSM called Business Driven IT Management (BDIM), which stresses business alignment in making IT management decisions. Claudio Bartolini summarised the BDIM approach as: “Bringing IT out of the technology provisioning domain means that one cannot consider it as an isolated entity and, in optimizing its performance, must also consider its complex interactions with the rest of the world” (Bartolini et al. 2011).

BDIM deals with complex relationships and must consider the irrationality of many human decisions as well as the more quantifiable metrics of traditional ITSM. BDIM is a relatively new approach to ITSM, but may be of more relevance to the e-Infrastructure sector than traditional approaches, as the ability to cope with unusual situations may be very useful in coping with the complexities of the e-Infrastructures.

2. e-Infrastructures

In the commercial IT service sector, SLM and SDM are clearly mature areas. However, these systems have rarely been used in the e-Infrastructure domain beyond small-scale pilots focusing on the activities within a single site. Where ITSM or SLM have been attempted, they have generally been tightly focused on development of Service Level Agreements, often ignoring the larger context or many other steps in the service management ecosystem.

The e-Infrastructure landscape covers quite a large range of different technology platforms, such as supercomputers (also called High Performance Computers or HPC), grids and clouds (sometimes referred to as High Throughput Computing or HTC), as well as data infrastructures and networked sensor systems. In Europe, the area has been called e-Infrastructure, partly due to the vocabulary of the European Commission that has funded the development of several aspects of it. In the United States, the most common term used for these services is Cyber-infrastruture, and funding has come largely from the Department of Energy and National Science Foundation—which includes the Office of Cyberinfrastruture (OCI).

2.1 Introducing the e-Infrastructure

In Europe, e-Infrastructure is considered to be made up from a number of levels (see fig. 9.5).

Whatever term is used, these systems have generally grown out of the academic domain and experimental systems developed in universities and research centres. Many of the solutions had initially a very tight link with specific research application groups whose process is characterised by globally distributed collaborations (such us High-Energy Physics, Earth Observation, biosciences and other application domains presented in, for example, the e-IRG Blue Paper (e-Infrastructure Reflection Group 2010), but have subsequently evolved into sets of multi- or general-purpose solutions and services. This means that while the levels in fig. 9.5 may seem to be independent to some extent, in reality the picture is more complex. Especially in its early days, many e-Infrastructure constituents were vertically integrated which, while sometimes desirable in a commercial context, was and has been a complicating fact in the efforts towards a more generic, multipurpose infrastructure.

Fig. 9.5. e-Infrastructure layers

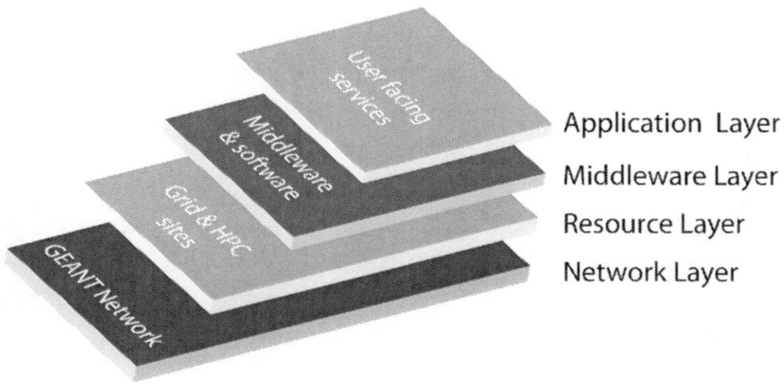

A fairly broad but comprehensive definition provided by the European Commission Seventh Framework Programme Research Infrastructures unit talks about:

> … empowering researchers with an easy and controlled online access to facilities, resources and collaboration tools, bringing to them the power of ICT for computing, connectivity, storage and instrumentation. This allows for instant access to data and remote instruments, "in silico" experimentation, as well as the setup of virtual research communities (i.e. research collaborations formed across geographical, disciplinary and organisational boundaries)
>
> (http://cordis.europa.eu/fp7/ict/e-infrastructure)

This highlights the nature of resource federation as something that is already inherent to the collaborations and core use cases (for example, federation is needed to realise remote access to instruments). At the same time the "instant, easy and controlled access" highlights the increasing demands for reliability and quality of service for the common services. It also argues against vertical integration, as it suggests a generic overall service, with portability of data, uniformity of access and simple comparison of providers.

For the moment, the largest federated e-Infrastructures are most likely the European HPC and grid services. The underlying European network,

GEANT is also of considerable size but focuses mostly on only one of the e-Infrastructure layers. Thus the HTC and HPC infrastructures are more interesting cases for ITSM analysis. For example, EGI (the European Grid Initiative http://www.egi.eu/) is the result of a series of EC funded projects (Enabling Grids for E-Science in Europe) that were funded for around €100 million over 6 years, and themselves resulted from a number of earlier European projects. EGI is intended to take on the infrastructure developed by these projects and convert the community and services they offered into a reliable, open and sustainable system. EGI acts as an umbrella organisation uniting a large number of National Grid Infrastructures (NGIs), national bodies (often formed as legal entities) that represent and integrate the grid computing sites in that country.

The overall EGI infrastructure consists of almost 350 resource providers from 57 different countries with hundreds of thousands of CPU cores (EGI 2011). It also connects to and works with other infrastructures around the world such as the Open Science Grid (http://www.opensciencegrid.org/), a comparable service in the United States. In comparison, the federated supercomputing infrastructure PRACE (http://www.prace-ri.eu/) consists of twenty-one European supercomputing centres that share resources to solve high-end HPC challenges. Similar US initiatives exist, such as TeraGrid, later replaced by XSEDE, the Extreme Science and Engineering Discovery Environment.

HPC and grid services both need to be provided in a reliable and predictable manner. They must also operate in situations where they are used in cross-disciplinary settings and outside the organisational structure of the research collaboration that developed them. Hence there is a need to formalise the working practices that have made the successful application possible so far, generalise them in a way that the services can be used without knowledge of the working practices of the original application domain and put into the context of the body of knowledge related to SLM and SDM in order to leverage mature tools and processes from the commercial IT management domain.

Grid infrastructures are an especially interesting topic of e-Infrastructure related research, as the minimum costs of establishing a grid site and participating in the overall infrastructure are considerably lower than with other e-Infrastructure components (for example HPC sites or shared, remotely accessible research instruments). During the decade of distributed service provision to the growing user communities, several organisational practices have evolved with relatively little organisational constraints. These practices, while often informal, have made it possible for the service providers to offer services that have satisfied the requirements of large, globally distributed user communities. Understanding the nature, benefits and limitations of these practices will be an important part in both successful application of a new generation of e-Infrastructure service solutions and in trying to generalise these collaboration models and apply them in other sectors of the society.

2.2 Clouds and e-Infrastructure

The arrival of the commercial cloud services had a profound impact on the e-Infrastructure landscape. Until the arrival of the cloud, e-Infrastructure offered the major distributed processing, storage and access solutions. Their relatively unchallenged positions meant that they were able to delay the development of clear sustainable business models and service management, as the lack of a commercial alternative allowed them to do so.

While clouds present slightly different usage scenarios to grids and are quite distant from HPC, they have nevertheless had a large impact on both. One clear effect was that clouds became a hyped technology, pulling commercial and media attention away from grids and HPC, among others. This could be seen in the ‘Hype Cycle’ diagrams provided by consulting and research form Gartner. Grid technology was already passed its peak of Hype in 2005 (Gartner Research 2005) but had disappeared from the cycle by 2008 (despite continued broad use in the scientific sector), while at the same time cloud computing began to peak.

Clouds also bring a clear commercial offering that offers a clear, simple metaphor describing a computing utility that is explicable to everyone from individual users to corporate chief information officers. Providing this simple service description is something that e-Infrastructure providers have struggled with. While HPC centres were able to make relatively straightforward offerings, the market for HPC was generally already aware of why they needed such high performance solutions and were able to cope with inherent complexities of the service descriptions. Grids, however, have always found it challenging to sufficiently simplify their service offering. In addition to the complexities of the different legacy software components and interfaces used to access grids, the challenge was related to the paradigm being based on resource sharing rather than purchasing. At present most of the users of European academic grids come with their own resources, or borrow unused ones from centres and other bodies willing to donate unused resources (with few, if any service guarantees).

Since their introduction, there have been efforts to meld clouds with existing e-Infrastructure. One interesting case is the StratusLab project discussed in the next chapter of this book. This EC funded initiative is attempting to introduce cloud and virtualisation resources into existing e-Infrastructure, notably to the European Grid Infrastructure.

2.3 Federated and aggregated infrastructures

Federation has been driven by a wide range of factors. These might be providing simple unified access to complex or heterogeneous resources, increased redundancy and improved reliability, or avoiding a lock in to a single solution. It can also stem from a need to composite resources to make a greater whole relatively cheaply, or cope with peak transient loads.

In fact most, if not all, of these motivations can be reduced in some sense to economic arguments. Existing resources might be brought together as it is not possible to buy new resources, or vendor lock-in may be seen as presenting unacceptable risks, due to dependence on outside organisations and their pricing policies.

In addition to economic factors, intangible "political" factors may also play a role. In grid computing, much of the initial need and technology came from the High Energy Physics sector and its requirements for computing to support the Large Hadron Collider (LHC), a particle accelerator at CERN. While funding was planned for the vast computing needs of the LHC, the funds were provided by numerous member states and participating institutes across the globe, all of them contributing to the cost of the system. Quite reasonably, these countries preferred to locate the resources, and the purchases, jobs and innovation potential that went with them, in their own countries. As a result, grid computing was used to allow the resources to be distributed, yet still act as a concerted whole.

The situation is not quite so clear in the cloud domain. Clouds bring together service units of computation or storage based on virtual machines (or groups of virtual machines) which when brought together form a single, coherent whole. The conceptual simplicity of the "ideal cloud," with freely distributable virtual machines, may make the hosting of the actual hardware seem a more commodity-like service. This service is less tightly coupled with the innovative aspects HTC service provision and use than the batch-oriented job submission systems, where a high-level job description creates an organic link between the "business case" of the user and the relative performance of a site. This allows direct comparison and benchmarking of sites in a way that is relevant to both service providers and the user communities.

Clouds are presented as federated infrastructures, but this is perhaps arguable. The difference between a federation and aggregation can be seen as a question of the autonomy of the constituent units. The autonomy of any single VM can be questioned. They are relatively “dumb” units that have a limited ability to cope with changing conditions on their own.

In contrast, traditional e-Infrastructure services bring together sites (grid sites or HPC centres), which have a high degree of autonomy and intelligence. Grids and HPC can scale by adding or removing sites assigned for the execution of the tasks, whereas clouds can expand or contract the pool of virtual machines. Clearly, this means that HPC and grid sites are more “expensive” service units than cloud virtual machines, but the larger and more complex basic units have been seen as an acceptable trade-off in cases where the tasks are inherently complex and on such a scale that a careful fine-tuning of the execution environment is seen as worth the effort. It also means that even if you generally only use a single access point (batch queue) to a grid, if it fails, other access points are relatively easy to identify. With a centrally managed cloud service this is not necessarily the case. In April 2011, Amazon’s Elastic Block Storage system failed (Dvorak 2011), apparently due to errors in the systems automated backup operations, locking accounts and taking down several major web sites. Such a central failure would be much more unlikely in a federated system but is quite possible in an aggregated one.

To use a metaphor, one might say that a cloud can provide as many inexpensive sand boxes as you like, each of which a few children can play in, but it can never provide a beach. Conversely, a grid converts (behind the scenes) many sandboxes into a beach, with the cost of additional complexity of having to provide lifeguards, parking lots and beach bars, most of which are either irrelevant or barriers of entry if the goal is just to allow a child to make few castles in the sand.

On the job submission level (“trip to the beach”), grids are able to scale quite effectively to new sites. They can do so because they are generally based on well-understood interfaces that can be adopted by any interested party and hide the complexities of heterogeneous resources. They are generally not centrally controlled. Even though some bodies provide some central monitoring for convenience, they are not required for the system to operate. The federation does not create additional single points of failure, and a site, group of sites or global region could fail without threatening the overall service.

It is also perhaps interesting to note that clouds appear to offer the suggestion of redundancy, without the reality of redundant services. Some commentators have noted that clouds actually show lower reliability and redundancy than the in-house systems they replaced. As one columnist noted:

> Never trust a single system component, be it a cloud, a network, a router, a database, or whatever … The typical solution is to fail to secondary components that can operate until the primary is back online. That used to be a given in IT. Unfortunately, many organizations have put too much trust into clouds, pushing their systems out to providers with the incorrect thought that a third party will provide the resiliency and the redundancy they require.
>
> (Linthicum 2011)

Perhaps the answer to this issue is to federate clouds themselves, as is the conceptual approach used in many of the multi-cloud initiatives. This involves spreading a service across multiple providers, whether they are internal (a so called hybrid cloud) or external. Such an approach requires vendor-neutral interfaces to multiple Infrastructure/Platform as a Service (IaaS/PaaS) services, something not necessarily simple to find or create consensus around.

Clearly, both clouds and traditional e-Infrastructure share a vision of simple and quick access to a scalable set of remote resources. These might be needed for one of the several reasons identified, whether a transient need for high capacity or increased redundancy. However, how these resources are brought together into a single unified whole varies from aggregated to federated approaches. While from the discussion here it may seem grids and HPC offer higher level of abstraction and functionality (at a higher cost), the differences in approach are also reflected in the service management, which will also have a major impact on customer satisfaction.

3. SLM-SDM in federated Infrastructures

3.1 Background to federated e-Infrastructure SLM

In the modern e-Infrastructure landscape, the academic community has developed most of the new distributed IT services. Despite the growing awareness of the need for guarantees of level of service, these services have been created and used in fairly informal environments, with tacit or "memorandum of understanding" type of agreements between the parties. This informality has undoubtedly simplified processes and lowered administrative barriers that might otherwise have impeded development. Had economic exchanges been introduced early in e-Infrastructure development for the academic sector, they might well have retarded the development of the overall services, especially when new technologies developed by individual teams had to justify their work to supervisors, introducing legally binding financial contracts (not only between multiple organisations, but organisations in different countries) would reasonably be a cause for concern. It would also be contrary to the general tendency toward informal collaboration in the academic sector.

Informal international collaboration is also seen especially strongly in the High Energy Physics community that strongly influenced grid development. The underlying economic issues of the collaboration were managed and understood, but their application domain was restricted to paying for major international equipment such as particle accelerators. Adding further economic agreements of smaller granularity would have been very difficult, with a large overhead related to consensus forming and administration in this particular environment.

However, as these infrastructures are maturing and are increasingly used in mission critical systems, their operation and management are becoming more formalised. However, this occurs against a background of service management aspects being developed at a much slower pace than the technical aspects of the service. This weakness has led to a number of initiatives funded by the European Commission to improve service management in e-Infrastructure. Projects such as SLA@SOI (http://sla-at-soi.eu/) have provided technical solutions for SLAs. On the other hand, the gSLM project supports improved formalisation of e-Infrastructure service management by bringing in knowledge from the commercial IT service management community.

3.2 SLM in the Cloud

When using commercial cloud offerings, the necessary contractual tools for IT service management are in place, but limited to a single cloud provider. The SLAs offered to customers are both well defined and weak, as the service guarantees are strong but the penalties for provider failure do not correspond well to the potential value lost by customers when services are unavailable. For instance, the SLA offered by Amazon for their EC2 cloud computing service promises an average annual uptime of 99.95%, which is promising if not the “five nines” (99.999) or “six sigma” service levels seen in some fields. The issue is that the maximum compensation offered by Amazon for breaking their SLA is a reduction on future purchases equal to 10% of the current contract (Amazon 2008). Compared to a major failure that may in extreme cases cost millions of Euros in reimbursements, penalties and lost revenues generated by the applications running on the cloud, a discount on future service purchases of the cloud itself is not proportionate to the losses.

Other cloud providers offer similar kinds of guarantee. LoadBalancer, a service from Rackspace, offers credit equivalent to 5% of the total fee for each hour the service fails to meet the SLA, and will provide credit up to 100% of the service purchase cost. Again however, this is credit against future purchases and ignores customer value generated on the cloud.

Service management on a multicloud scenario is more confused. At present, commercial multicloud offerings appear to not actually be federators or aggregators. Rather, they appear to be essentially dashboards for monitoring multiple cloud services from different providers. While this may allow, for example, monitoring of a service replicated across two commercial providers, it will not federate the services itself. Also, the SLAs remain between the customer and each of the clouds; the commercial multi-cloud service does not usually provide services and contractual tools that would allow linking them with client's SLM processes.

As an example, if a composite service has parts running on two clouds and one of the clouds failed, the multicloud provider would note this, but the customer would still pay for the working parts of the service, even though the system was no longer able to provide the composite service they purchased it for. Compensation in such a case moves even further out of balance with customer perception of value.

A clear issue here is that current SLAs offered by cloud providers don't offer the guarantees needed for federation of their services. While a company could in theory federate multiple commercial clouds and offer a service across them to a customer, they would have to price the composite service a great deal above the costs of the individual services to be able to offer strong service guarantees that would, in case of failure, pay compensation on behalf of the commercial cloud provider that failed. This would not currently be economic.

Multi-clouds in the academic domain, while truly federating cloud resources, are in the same position to SLM as academic grids and HPC: they have at the moment little if any formal SLM. They tend to run over open source cloud solutions, which do at least reduce potential for a lock-in with an individual provider. However, as the providers do not offer well-defined SLAs, the maturity of academic multi-clouds cannot be seen as being more advanced than the commercial clouds. The commercial and academic systems simply are on different maturity levels, depending on which aspect of the overall system is under observation.

3.3 Sharing lessons learned

Developing formal SLM-SDM models based on the day-to-day practices of delivering grid services may provide input for improving service provision in other domains. The nature of grids as federated (rather than aggregated or unitary) infrastructures can be considered quite unique, especially when taking the level of dynamic allocation of resources into account. Other emerging federated systems (such as multi-cloud environments) will face many of the same challenges as grids and can therefore learn from the challenges they have faced. The multi-cloud systems are perhaps the closest analogues to grids, as they must integrate resources from heterogonous providers in the same way, and the technical barriers of entry to the cloud provider market are similarly low.

Improved service provision, in grids, clouds or other e-Infrastructures also facilitates the assessment and comparison of costs and service quality. This is increasingly important in a era of reduced research budgets and a greater emphasis on users such as those represented by projects of the European Strategy Forum on Research Infrastructures (ESFRI) (http://ec.europa.eu/research/esfri), which represents major research communities with infrastructure needs. It seems likely that as time goes on, users and projects will be asked to demonstrate the efficiency and cost effectiveness of the infrastructures they use (and fund, directly or indirectly), and SLM-SDM will play an important role here.

3.4 Federations of users

An area where the e-Infrastructure domain differs from the mainstream cloud visions is the federation of users into cross-organisational Virtual Organisations (VOs) or Virtual Research Communities (VRCs). This practice stems from the collaborative research ventures that may see hundreds of institutes participating in the building of a single physical research apparatus.

More recently, as more and more of the new research infrastructures are either virtual or multi-site in nature, this model has become more and more commonplace. Usually, the authentication of a user is still tied to a specific institute related to the physical location of the person. However, the authorisation decisions are tied to the membership of this user in a specific, cross-organisational group. This aspect of a federation is one of the issues that challenge the established SLM/SDM principles at a fundamental level.

While at the moment the practice of federating users seems to be limited to the grid/HPC infrastructures, one could imagine situations where a group of collaborating companies (for example in shipping or construction domain) could create project-based groupings for their employees and leverage their combined negotiating power for ICT services while retaining complete autonomy in terms of authorisation technologies and tools.

3. 5 Modelling relationships

One of the key differences between traditional IT Service Management and that emerging in federated computing ecosystems is based on the complexity of the relationship model. The number, type and structure of the relationships inherent in the system are much higher in the federated computing ecosystems than in the traditional ones.

Traditional ITSM relies on a fairly simple and linear, or at least hierarchical arrangement of organisations involved in a supply chain. Each relationship can be mapped to a fairly clear group, a customer, a provider, and a sub-supplier, see fig. 9.6 below). Each relationship within this is bilateral, unique and captured in a legally enforceable agreement. As a result it is easy to identify points of failure and therefore assign responsibility for service failure.

Commercial, single clouds actually follow this schema, but the contract and the enforcement of the SLA that captures the relationship between provider and customer is not up to the ITSM standards. We expect this to change as customers, especially commercial ones, increasing realise their cloud services do not provide their users the same standards they demand in other IT service suppliers. ITIL or ISO/IEC 20000 compliant cloud services might show increased cost, but would also provide or support more value for customers.

Multiclouds show two models today, one commercial and one academic. While both can in theory be reduced to simple, linear hierarchies, both types fail to stand up to rigorous SLM, as shown by fig. 9.7 below.

The commercial alternative has too many relationships, as the multicloud provider is actually providing a monitoring service not a cloud service. This means the customer takes on the risks of federated clouds, as the guarantees provided by the cloud broker are not sufficient. On the other hand, the academic multicloud is well structured (i.e. it doesn't add complexity) but doesn't use sufficiently strong agreements between the actors since they lack clear, enforceable SLAs as components.

Grids show a much more complicated arrangement than the cloud scenarios due to the very large number of organisations involved in provision of services through academic grids.

Fig. 9.6. A simple supplier chain scenario

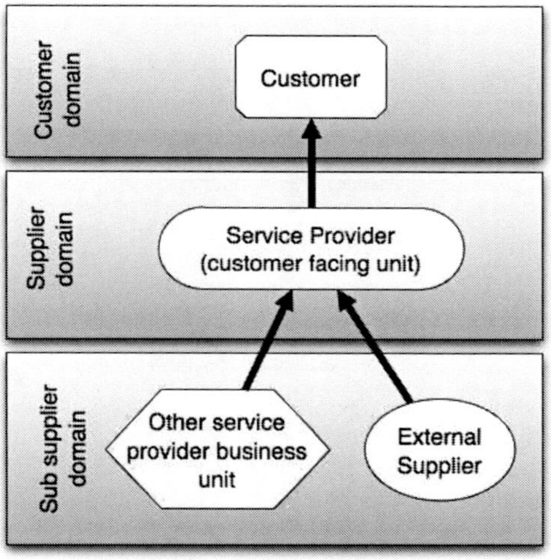

Fig. 9.7. Two kinds of multicloud—commercial and academic

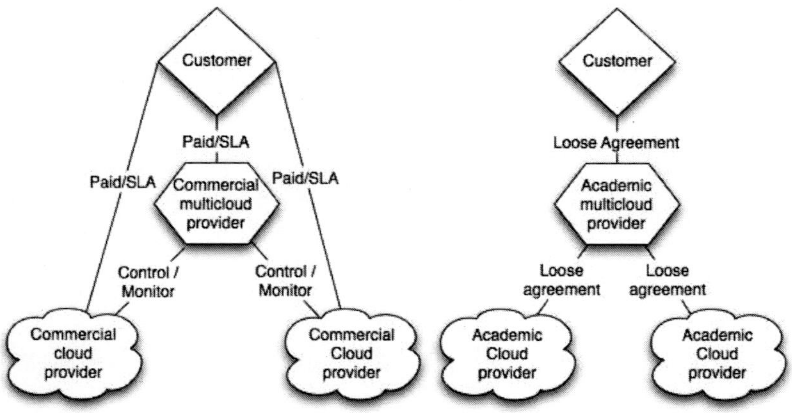

One of the clear needs in the grid ecosystem is to reduce this complexity to something simple enough to be captured in a system of SLAs and contracts. Reducing complexity and formalising the agreements between the key stakeholders should be seen as prerequisites for applying and extending commercial ITSM systems such as ITIL to cover grid-like federated system.

Fig. 9.8. Grid relationships: many lateral and non-hierarchical relationships make the relationship map complex and make SLM implementation more difficult

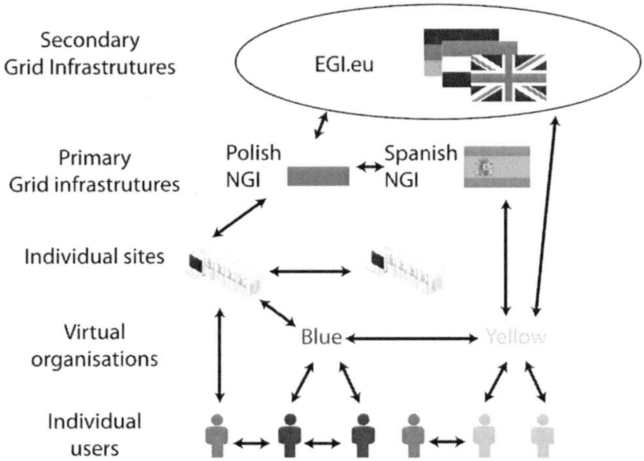

4. gSLM project approach

The gSLM project works on the areas described above, seeking to bring experiences and approaches from commercial ITSM and provide ways to provide them in the grid and e-Infrastructure domain. It is a two-year support project including experts in grid operations as well as formal service level management. The project operates through internal and external workshops, consultation with experts and discussion and collaboration with a range of other projects and initiatives, including EGI, PRACE, IGE (http://www.ige-project.eu) and mOSAIC (http://www.mosaic-cloud.eu/).

By combining input from these different sources with expertise from the project consortium, gSLM provides recommendations at both policy and practical levels, aimed at e-Infrastructure operators and users. The project is not yet complete, but the following sections briefly describe its approach, and current and planned outputs.

4.1. Establish common terminology and model

The e-Infrastructure and SLM-SDM fields have notably different terminologies, often using the same terms for different things. By creating a common terminology, gSLM facilitates communication and translation of concepts and approaches between the two fields, and experts from both communities have participated in developing and improving a common glossary.

In parallel to the analysis of the terminologies, the project is setting out an ontology and a set of use cases to represent the main relationships and processes in provision of grid services. These will be used to create a two-part model of grid service provision. The first part will consider the current model of grid relationships, while the second will consider a more idealised model that will facilitate improvement of grid service provision, and also be of interest to other federated infrastructures. The idealised model is presented in the fig. 9.9.

Focusing the effort of formalisation of the service level and delivery management efforts on the "necessary" relationships should provide the maximum return on investment for the efforts. The relationships labelled "unnecessary" may provide additional flexibility in the testing or piloting phases, but in the production environments they should be considered exceptions rather than relationships that are covered by the ITSM processes.

Fig. 9.9. Optimised grid relationship model

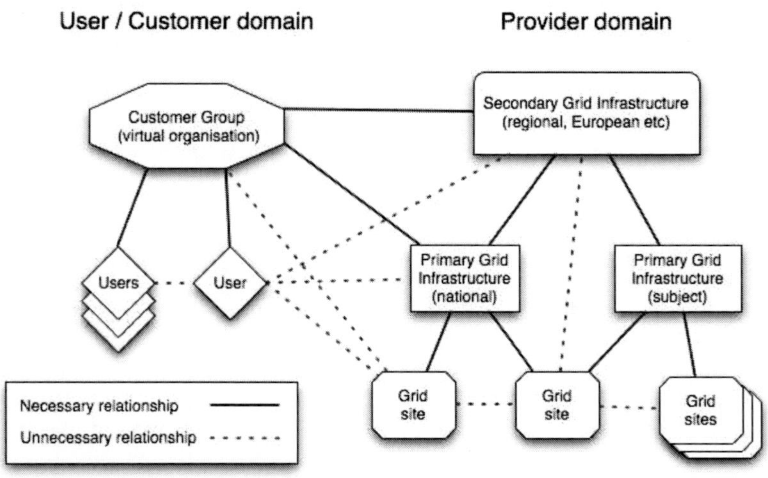

4.2. Study the current approaches and the impact of incremental formalisation

Building on the terminology and model, gSLM produces supporting material and studies on SLM-SDM in e-Infrastructures. The initial focus will be on the importance of SLM in grids, but it will be subsequently extended to a requirements catalogue for SLM-SDM in grids. The task takes input from a variety of sources, including the gSLM grid model and user needs that lead from them. It also draws on surveys, consultations at gSLM events and direct contacts with experts. The papers also draw on analyses of grid theory to identify the limitations of the current implementations of grids with their minimal SLM-SDM.

The results of the case studies provide backdrop and hypothetical test scenarios that can be used to predict the potential impact of the recommendations gSLM makes.

4.3. Create maturity model and roadmap

The final outcome of gSLM is a roadmap on improving SLM in grids and federated e-Infrastructures. This will draw together the earlier outputs of the project as well as providing tools, techniques and approaches that allow all members of the grid and broader e-Infrastructure community to improve service provision at both the practical and policy levels.

After recapitulating the terminology and grid SLM model requirements and presenting a summary of the white paper on the importance of grid SLM, the roadmap will describe a maturity model for SLM in grids that has been developed based on consultation with the community, expert opinions and case studies. Recognising that the maturity of an SLM system needs to match the maturity of the infrastructure it serves, this describes appropriate levels of service management through the evolution from an experimental service to an infrastructure in use by a global community.

This model uses levels based on maturity models in the COBIT ITSM framework. The short descriptions of the maturity levels are as follows:

- Level 0 Non-existent: There is no consideration of service management at all.
- Level 1 Initial/ad hoc: There is awareness that service quality has to be addressed.
- Level 2 Repeatable but intuitive: There is some initial, limited formalism in the provision of the system.

- Level 3 Defined: Formal contracts exist, as does some sort of service catalogue, but rules are not always followed.
- Level 4 Managed and measurable: Contracts are not only formalised but also effectively monitored and enforced.
- Level 5 Optimized/Integrated: The management of the system has gone beyond formalism and repeatability into optimisation of the system.

Following this, the roadmap will provide general guidance and steps for deploying improved SLM, divided into the areas in which they must be applied: Policy & governance, technology, organisational & process change and lastly human & social factors.

These general guidelines will then be used to produce specific prescriptions for different types of actor in the grid community, showing how users, sites and infrastructures can all take steps to improve service provision.

Finally, the roadmap will consider the impact that adopting the maturity model and recommendations could have on current infrastructures. It will also suggest an approach for future work in the area.

4.4 Summary of initial findings of the gSLM project

The federated nature of grids means that many relationships that would be described in OLAs in industry, such as between data centers in a single cloud provision organisation, are actually between sites that are distinct legal entities, making them closer to UCs. UCs will generally have to be stricter or more formalised than OLAs, as they will be between organisations, yet in the grid sector they remain quite informal. As far as the SDM use cases are concerned, we have so far identified seven categories of processes that are relevant to grids. These process categories include, among others, service configuration, service availability, service continuity and service capacity.

In addition to analysing the types of contracts that would capture the dynamics of the grid infrastructures, it is necessary to create groups of actors within grids that can be used for modelling purposes, as described in previous sections. A simple schema for these could be:

- Virtual Organisations (users/clients)
- Grid Infrastructures (GIs—used as a common name for varying levels of infrastructure). These can include:
    - National Grid Infrastructures or other primary groupings

    - International and other secondary Grid Infrastructures, such as EGI.
- Sites
- External suppliers

Complication arises from the realisation that despite the existence of many examples of each of these groups, in fact the nature of organisations that would put themselves in one of these groups is not consistent. In particular, what a National Grid Infrastructure actually is has not been agreed. Equally, the services that EGI will provide are not yet fully clear. This complicates the structuring of the community in a rational way, so the gSLM project hopes to engage with these groups to help them make rational and sustainable decisions about the function of each actor class.

The maturity model may play a role here, as it gives a somewhat objective tool for looking at SLM maturity, and in fact the overall maturity of entities in this ecosystem. It also lets gSLM more easily take the steps needed by each actor to improve their SLM and overall. This originates from the realisation that experimental services cannot be set in stone and burdened by excessive formalisation, while it is unwise to build services designed to scale to many thousands of users on a basis of informal processes with no consequence for service failures, leading to unhappy (or no) users. One aspect of the maturity model is dividing the existing *de facto* contractual relationships to ones that can be formalised efficiently and that should be the focus of the SLM-SDM efforts (for example "VO-GI" or "GI-higher level GI"), and the ones that should be considered only in the experimental stage (for example "user-site" or "VO-site").

5. Conclusions

The multi-cloud vision of dynamic service composition and re-composition based on a pay-as-you-go model, on a structure mostly based on bilateral service contracts between the providers and consumer, does seem some distance from the traditional e-Infrastructures, especially the NGI/site hierarchy in grids which may seem to restrict the applicability of the gSLM results to the software and services domain. However, the decision to discard the lessons learned from the decade of grid service provision should not be taken lightly, especially since the e-Infrastructure projects are integrating cloud interfaces and services as a component in their offering.

Despite this initial appearance of fundamentally different structures between grids and multi-clouds, gSLM relies on a number of use cases that describe service-level management requirements common to IT service provision that are independent of the actual middleware or provider consumer relationship. Moreover, the model adheres to industry-standard, widely used concepts and specifications like ITIL or ISO 20000. Therefore, we are confident that the SLM-SDM model defined by gSLM carries commonalities that also apply to service-oriented ecosystems.

We would also assume that once the demand for stronger SLA guarantees grows on the commercial side, the providers will have an interest in building arrangements that allow them to honour their service level agreements also in cases where their own infrastructure fails. In effect, the commercial world will need to move from monitoring or aggregation towards genuine federation of resources from multiple providers. The agreement structure enabling this federation of resources may well bear resemblance to the SLM/SDM structures modelled by the gSLM project.

In particular, it would be interesting to see how to map the roles of the sites, NGIs, and EGI VO to the federated multi-cloud ecosystem. For example, the mapping the customer group/Virtual Organisation may simply be based on the authentication/authorisation aspects, but it will also change the balance of power in contractual disputes. A VO representing users or a credit card company handling billing information of the clients has the ability to influence the behaviour of the group of providers more than a single user.

For example, one could imagine credit card companies becoming the channel for the cloud users to deal with SLA failures by requesting charge-back on the card payments to cloud providers, for example based on local laws that do not allow limiting implied warranties. Even a low volume of such transactions would provide a strong economic incentive for credit card companies to use alternative dispute resolution models, as the administrative overhead of dealing with this kind of dispute is very high for all parties (except the customer making the complaint). The role the credit card company would play might end up having several commonalities with the role of the "idealised VO" in the gSLM model. In this scenario, the cloud providers might also find it reasonable to establish a stronger federated model to deal with the risks of increased transaction costs or to manage increased liability risks.

Parallel to the maturing of the commercial cloud and multi-cloud ecosystems, there are technology, governance and policy related developments that are significantly changing the e-Infrastructure landscape. Hence, studying the organisational structures and processes in a way that makes transferring the ITSM-related results and best practices between grids, (multi-) clouds and HPC services is a particularly timely topic. ITSM and BDIM are becoming more and more important for academia, traditional (commercial) IT and the emerging multi-cloud environments as they face new budgetary constraints, more and more globalised competition while at the same time trying to maximise the benefits of the new technologies. While the level of technical complexity—at least as experienced by an average user—is constantly being reduced, the way organisations deal with legal, administrative and managerial aspects of IT service provision are becoming areas in which they should focus their efforts when trying to differentiate themselves from the competition.

Acknowledgements

The authors would like to thank the entire gSLM consortium for contributions that underlie this chapter. The gSLM project (Service Delivery and Service Level Management in grid Infrastructures) is co funded by the European Commission under contract number 261547, as part of the EC's Research Infrastructures programme. More information on gSLM can be found on the product website at http://www.gslm.eu.

References

Amazon "Amazon EC2 Service Level Agreement", 2008. http://aws. amazon.com/ec2-sla/ - accessed 3.4.2011

Bartolini, C., Stefanelli, C. and Tortonesi, M. "On Decision Making in Business Driven IT Management." In proceedings of the 6th IFIP/IEEE International Workshop on Business Driven IT Management (BDIM2011) Dublin, Ireland, 2008.

Cartlidge Alison and Lillycrop Mark (Editors). "An introductory Overview of ITIL V3," 2007. http://www.best-management-practice.com/gempdf/itSMF_An_Introductory_Overview_of_ITIL_V3 .pdf (accessed June 30, 2011)

Dvorak John C. "The Amazon Cloud Failure Redux", PC Magazine, 2011. http://www.pcmag.com/article2/0,2817,2384274,00.asp (accessed July 22, 2011)

e-Infrastructure Reflection Group. "e-IRG Blue Paper" 2010. http://www. e-irg.eu/images/stories/eirg_bluepaper2010_final.pdf (accessed July 14, 2011)

EGI. "Figures and utilization," 2011. http://www.egi.eu/infrastructure/ Figures_and_utilisation/ (accessed July 1, 2011)

Efficiency and Reform Group. 2011, on Cabinet Office website http://www.cabinetoffice.gov.uk/unit/efficiency-and-reform-group (accessed August 9, 2011)

Gartner Research. "Gartner's Hype Cycle Special Report for 2005," http://www.gartner.com/resources/130100/130115/gartners_hype_c.pdf (accessed August 2, 2011)

ISO. Homepage of the standard, 2011. http://www.iso.org/iso/iso_catalogue/catalogue_tc/catalogue_detail.htm?csnumber=51986 (accessed August 1, 2011)

Linthicum David. "The failure behind the Amazon outage isn't just Amazon's", InforWorld, 2011. From http://www.infoworld.com/d/cloud-computing/the-failure-behind-the-amazon-outage-isnt-just-amazons-107 (accessed May 6, 2011)

PART IV:

CLOUD ADOPTION

CHAPTER TEN

STRATUSLAB CLOUD DISTRIBUTION

CHARLES LOOMIS, MOHAMMED AIRAJ,
MARC-ELIAN BÉGIN, EVANGELOS FLOROS,
STUART KENNY AND DAVID O'CALLAGHAN

Cloud technologies provide many benefits for scientific and engineering applications, such as customised execution environments, near-instantaneous provisioning, elasticity, and the ability to run user-level services. However, a rapid, wholesale shift to using public, commercial cloud services is unlikely because of capital investments in existing resources and data management issues. To take full advantage of cloud technologies in the short term, institutes and companies must be able to deploy their own cloud infrastructures. The StratusLab project provides a complete, open-source cloud distribution that permits them to do this. The StratusLab services include the computing, storage, and networking services required for an Infrastructure as a Service (IaaS) cloud. It also includes high-level services like the Marketplace that facilitates the sharing of machine images and Claudia that allows the deployment and management of complete software systems.

1. Introduction

Despite the enormous marketing hype around cloud technologies, much of the excitement is well deserved. Cloud infrastructures provide users with more agility by avoiding lengthy procurement delays, access to larger computing resources by avoiding constraints from local physical infrastructures, customised execution environments, and the possibility to expose user-level services rather than just static content.

Cloud computing has been embraced by small and medium-sized enterprises (SME) as a means to limit capital expenditures and to allow for fast, even exponential growth. Those in the academic sector have also taken notice and are investigating how cloud computing can provide results more quickly and at lower cost. This is especially true for academic areas that traditionally lack experience with deploying and maintaining local computing infrastructures, like the humanities and social sciences.

The desire to maximise elasticity at minimum cost pushes public cloud providers to deploy massive data centers in locations where energy and cooling costs are low. Environmental concerns force providers to consider also the carbon footprint of their operations. Computing infrastructures are swinging back to the centralised model of the mainframe era.

However, several factors slow this movement towards a centralised computing model based on public cloud infrastructures. Inertia is one factor, where existing investments in physical (and human) infrastructure mean that current resources cannot be immediately “outsourced” to a cloud. User confidence in the cloud is another. Users must be convinced that cloud provides a reliable, robust service, especially in the face of recent large-scale outages of Amazon’s cloud (Svensson 2011).

Data management, however, is the most fundamental impediment. On the commercial side, privacy, confidentiality, and legal constraints hinder remote storage of data, although some commercial offerings are now appearing to address this issue. On the academic side, data are generated by a variety of instruments, are naturally distributed, and need co-located computing resources. Because of this, there will be a need for local computing resources at companies and at universities for some time to come.

Companies and universities still want to take advantage of the numerous benefits of cloud technologies. To do so in the short term, they must be able to deploy a cloud in addition to or instead of using a commercial cloud provider. The StratusLab project offers a complete, open-source, Infrastructure-as-a-Service (IaaS) cloud distribution to satisfy this need.

2. StratusLab Project

The StratusLab project (http://stratuslab.eu/) funded in part by the European Commission, aims to develop a complete, open-source cloud distribution that can be deployed in production in both the academic and commercial sectors. It consists of six partners from five European countries (France, Greece, Ireland, Spain, and Switzerland) and runs for two years, ending in May 2012. All of the partners have previous experience with virtualisation and with distributed computing, particularly grid computing.

2.1 Cloud Technology

The term "cloud," as used in the press and popular literature, describes a wide range of technologies and deployment strategies. Marketing deliberately misuses the term to take advantage of the current excitement surrounding the technology, and further muddies the water in doing so. NIST, fortunately, has published excellent definitions of the different types of cloud service and deployment models (Mell and Grance 2010). Table 10.1 below defines the different cloud service models, based on this publication. The cloud service models form a rough hierarchy of services, where platforms are built on IaaS clouds and services are built on PaaS clouds.

There are also a variety of deployment models. Private clouds are run by an organisation for its own users whereas public clouds are run by a provider for the general public. These have significantly different security requirements although the underlying cloud technologies are the same. A community cloud is an intermediate deployment between private and public clouds providing a collaborative infrastructure involving users and possibly providers at different institutes. Hybrid cloud infrastructures bridge different cloud deployments to federate the available resources.

Table 10.1. Cloud Service Models

Software as a Service (SaaS)	Consumers use the provider's applications that run on a cloud infrastructure. The applications are accessible from various client devices through a thin client interface.
Platform as a Service (PaaS)	Consumers deploy onto the cloud infrastructure (created or acquired) applications developed using programming languages and tools supported by the provider.
Infrastructure as a Service (IaaS)	Consumers provision processing, storage, networks, and other fundamental computing resources where the consumer is able to deploy and run arbitrary software, which can include operating systems and applications.

The underlying ideas of clouds are not new, having their roots in the commodity and utility computing of the past two decades. Those initiatives, however, had limited success compared to cloud computing. The popularity of cloud computing comes from the convergence of three factors:

- Mature virtualisation technologies imposing very small performance penalties compared to bare-metal performance,
- Simplified APIs and usage models, particularly RESTful (Fielding 2000) APIs running over the standard HTTP(S) protocol, and
- Excess commercial computing capacity.

This last point is especially important as it allowed cloud computing to be made available to a wider audience at a reasonable price.

The StratusLab cloud distribution allows an IaaS-type cloud to be deployed; it also provides functionality to support PaaS cloud installations over the IaaS cloud. The distribution supports any type of cloud deployment: private, community, or public clouds. Future versions will support federation of cloud infrastructures into hybrid clouds.

2.2 Existing Software

Many commercial software vendors now provide solutions for deploying private clouds. Probably the best known is the ESXi suite from VMware. Most academic institutes and smaller companies cannot afford these solutions because the costs can equal or exceed the costs for the hardware itself. They also tend to be less adaptable as users do not have access to the code to make changes for their environment or to experiment with alternate scheduling/provisioning algorithms.

At the time the StratusLab project was proposed, there were no complete, open source cloud solutions available. However, there were many toolkits for virtual machine management. The best-known toolkits are Nimbus (http://www.nimbusproject.org/), Eucalyptus (http://www.eucalyptus.com/) and OpenNebula (http://opennebula.org/). Since that time, all of these toolkits have become more functional and provide a larger set of the services necessary for creating an IaaS cloud.

OpenStack (http://www.openstack.org/), another initiative started concurrently with Stratuslab, also aims to provide an open source IaaS cloud distribution. The planned services of OpenStack mirror closely the compute, storage, and networking services in the StratusLab distribution, but are accessible via different APIs. Future evolutions of both platforms will hopefully move toward standard APIs and provide some level of interoperability for users of the two systems.

Because of the desire to provide a production quality cloud distribution quickly, one of StratusLab's core principles is to reuse and build upon existing software when possible, for example OpenNebula and Claudia. The project choose to use OpenNebula for virtual machine management because of its maturity, open architecture permitting extensions, and use of a software license (Apache 2) compatible with commercial reuse of the StratusLab results. Claudia permits the deployment and control of complex, multi-machine services, and consequently is a critical component for supporting "Platform as a Service" (PaaS) infrastructures above StratusLab clouds. Both of these packages emerged from the RESERVOIR (http://reservoir-project.eu/) research project and have since been extended and hardened for use within StratusLab.

3. Existing e-Infrastructures

European academics have access to a wide variety of existing infrastructures, ranging from local clusters, through university computing centres, to large European computing platforms. An important goal of the project is to ensure that academics at all levels can take advantage of cloud computing through the StratusLab cloud distribution, while maintaining its relevance also in the commercial sector.

3.1 Commercial

Cloud technology really was born in the commercial sector, stemming initially from Amazon's efforts to monetise their excess computing capacity through an IaaS offering. Google with Google App Engine and Microsoft with Azure have since also provided public cloud infrastructures at the PaaS level.

All of these platforms have been popular with start-ups and small enterprises, allowing them to avoid capital expenditures on computing hardware and offering the possibility of quick expansion as the business grows. They have had more limited success for larger enterprises and for academic users, mainly because of data management issues. For commercial entities, issues concern privacy and confidentiality of sensitive data. For academics, the issues revolve around the accessibility of data and its large size.

An open source cloud distribution makes cloud technologies more accessible to larger enterprises by allowing private cloud deployments that overcome the data protection issues. Similarly, cloud deployments (either public or private) near large data stores ensure fast, reliable access to academic data.

3.2 PRACE

The Partnership for Advanced Computing in Europe (PRACE) (http://www.prace-project.eu/) is an initiative to provide a world-class supercomputing service for European research and industrial users. The service will target users with parallel applications that need very high CPU performance and inter-process communication bandwidth.

These highly optimised applications are unlikely to benefit from cloud computing because of the slight degradation in performance due to virtualisation and because most cloud infrastructures cannot provide high-performance network interconnects. Consequently, this supercomputing infrastructure will remain an important complement to other computing platforms, such as cloud and grid platforms, based on off-the-shelf hardware.

3.3 EGI

The European Grid Infrastructure (EGI) (http://www.egi.eu/) provides a high-throughput computing platform with resources distributed throughout Europe. This platform is open to European researchers or those researchers associated with a European project. The infrastructure itself comprises more than 300,000 CPU cores and 100 PB of disk space, contributed by approximately 350 participating computing centres in 57 countries.

The infrastructure has been very successful in supporting scientific endeavors with over 13,000 users taking advantage of the infrastructure's resources. Although open to all scientific disciplines, high-energy physicists, specifically researchers associated with the Large Hadron Collider at CERN, are by far the largest contributor and consumer of resources available through the infrastructure.

For historical reasons, the primary grid middleware used on the infrastructure is gLite (http://glite.org/), although there are on-going efforts to support also UNICORE (http://www.unicore.eu/), ARC (http://www.nordugrid.org/arc/), and Globus (http://www.globus.org/). All use a model based on job submission and appear, crudely, as distributed batch systems. This model works well for applications or calculations that can be easily parallelised into large numbers of independent tasks, like those in high-energy physics.

However, many other applications need customised execution environments or involve the deployment of user-level services and are not well adapted to the computing model of the grid. StratusLab, working with EGI, aims to bring cloud technology to this platform to make the infrastructure more flexible and, hence, more attractive to a wider variety of users.

4. Combining Grid and Cloud Technologies

Some view grid and cloud technologies as competing technologies; this project however views them as complementary. Combining those technologies promotes scientific e-infrastructures that are more dynamic, appeal to a wider variety of scientific communities, and encourage international and interdisciplinary collaboration.

Models to combine these technologies can be crudely categorised as "grid-over-cloud" or "cloud-over-grid." The cloud-over-grid approach uses the existing grid job submission and management services to launch (customised) images for job execution. The CNAF data center in Italy has used this approach for their WNoDeS (http://web.infn.it/wnodes/) service. This does provide a customised environment for users, but still presents new users with the high learning curve of grid services and still maintains a "job-based" workflow that is not applicable to all applications.

StratusLab takes the opposite approach—"grid-over-cloud"—specifically treating the grid as a Platform-as-a-Service (PaaS) infrastructure running over StratusLab's Infrastructure-as-a-Service (IaaS) cloud (see fig. 10.1 below). This architecture allows system administrators to expose their resources via both grid and cloud interfaces to their users, maximising the site's utility. By offering both interfaces, it allows users to decide which abstraction is best suited to their applications—a job-based abstraction for the grid or a machine-based abstraction for the cloud. Users can exploit fully the dynamic nature of the cloud, provisioning resources as needed and running user-level (and community-level) services as appropriate.

However, the layering of grid services over the cloud does require some detailed integration to ensure that they interoperate well, especially from the user's point-of-view. For example, the grid and cloud interfaces should support the same authentication and authorisation mechanisms to ensure that users can take advantage of both types of services and potentially mix them for a particular calculation.

Fig. 10.1. Grid as a PaaS

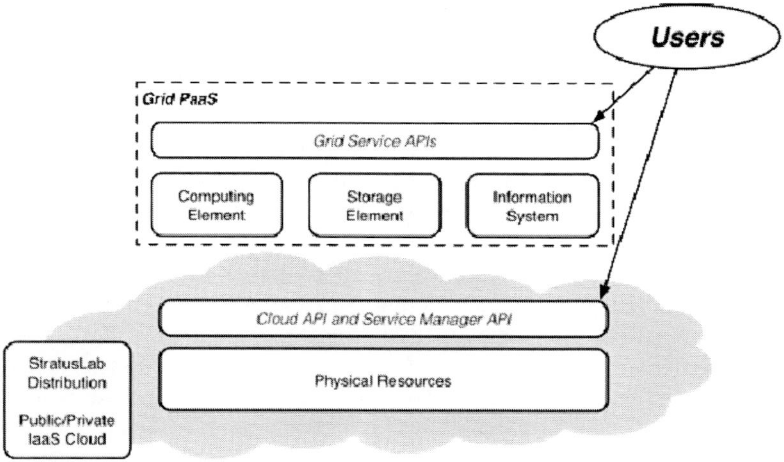

To test this scenario and ensure that it works, StratusLab runs a production grid site within EGI. This site has been certified via the standard EGI procedures and is monitored and evaluated with the standard tools. This experience has shown that StratusLab can indeed support running grid services within a cloud, although direct use by scientists at this point has been limited. With the production release of the StratusLab distribution, the project expects pilot sites to adopt this architecture, providing further information on running this scenario and allowing the project to further refine the distribution.

5. Requirements

At a minimum, the StratusLab cloud distribution must provide all of the services for a complete IaaS cloud deployment, including compute, storage, and networking services. The compute services must allow the deployment of virtual machines created by users themselves and should pose no constraints on the type of operating system used within the virtual machine. Similarly, a standard, lightweight contextualisation method must be used to minimise perturbations to the user's standard machine configuration. The storage services must allow persistent storage of data with a lifecycle independent from a particular virtual machine instance. The system must allow services running on virtual machine instances to be accessible from the wide area network, if requested by the user.

The StratusLab distribution must also support operation in a distributed, collaborative environment like that of EGI. This requires a flexible authentication framework that allows integration with grid authentication mechanisms as well as existing authentication mechanisms at sites deploying standalone cloud infrastructures. The distribution must also be designed to foster collaboration by, for instance, allowing the easy sharing of virtual machine images. Many scientific domains will build their own computing platforms above an IaaS cloud; consequently, the distribution must provide services to facilitate the deployment and operation of these domain-specific PaaS.

StratusLab targets a wide range of data centers, ranging in size from a few tens of machines to tens of thousands. Consequently, the services must be scalable. The experience of the system administrators at the targeted sites also varies greatly. As a result, the distribution should be easy to install but also be compatible with the automated installation tools of larger sites. Finally, different sites prefer different operating systems, so the distribution should demonstrate portability between several operating systems.

6. Distribution

The project uses agile software development methodologies to ensure that the StratusLab cloud distribution meets the requirements of our users and evolves quickly while maintaining its production quality. The project always moves from working version to working version, making iterative improvements and testing continuously, ensuring that all versions can potentially be released.

Because of this methodology, StratusLab released at the end of its first year the 1.0 production distribution, preceded by four public beta releases. The first beta release, three months into the project, was already a functional cloud; the project has been running a reference cloud infrastructure with its software since the second beta release.

A StratusLab cloud based on the 1.0 production release offers computing, storage, and networking services as well as high-level services to facilitate sharing of machine images and deployment of multi-node services (see fig. 10.2 below). For users, the distribution also provides a set of client commands for interacting with the cloud services. These commands are written in python for portability, running on Linux variants, Mac OS X, and Windows. For system administrators, deployment can be done via a scripted "manual" installation or via a fully automated configuration with the Quattor (http://quattor.org/) fabric management system.

Fig. 10.2. StratusLab Architecture

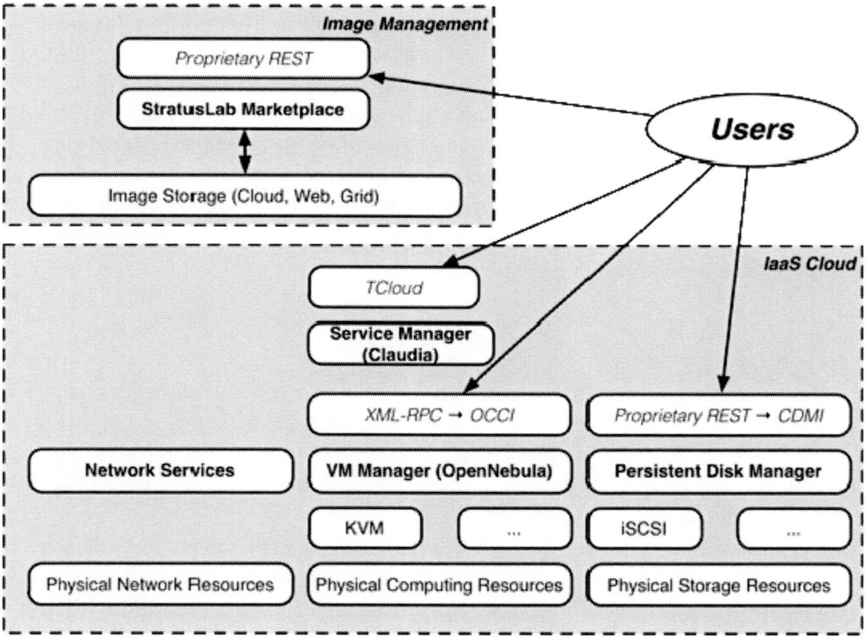

6.1 Computing

Virtual machines are at the centre of any IaaS cloud, and consequently virtual machine management is a core feature of any IaaS cloud distribution.

The StratusLab software stack for computing services consists of the operating system running on the physical hardware, a hypervisor for local VM control, a virtual machine manager (VMM) for cloud-wide management of VMs, and a cloud API permitting remote access to the resources. Currently, StratusLab supports Fedora14 at the OS-level, uses KVM for the hypervisor, integrates OpenNebula as the VMM, and exposes OpenNebula's proprietary XML-RPC interface as a cloud API.

OpenNebula allows users to control the complete lifecycle of a virtual machine from its initial creation to ultimate destruction. It provides many hooks into the machine lifecycle and various customisation points that allow it to integrate well with the other StratusLab services.

The project has also worked with the developers to improve the system. To help with forensic analysis following security incidents, a quarantine system was introduced and more detailed logging is provided. Other improvements include better management of network addresses, better error feedback to users of the cloud API, and eventually user, group, and role support in the authorisation system.

Future versions of StratusLab will take better advantage of the flexibility of OpenNebula, allowing the support of more underlying operating systems and more hypervisors. The project will also move to OCCI (http://occi-wg.org/), a proposed OGF standard, for the cloud API, potentially allowing users a common API for different cloud implementations.

6.2 Storage

All non-trivial analyses use input data for a calculation and store its result. Consequently, storage resources are a core component of any computing infrastructure. StratusLab uses a "disk-based" abstraction for its storage services providing tools to manage and to exploit both static (read-only) disks and persistent (read-write) disks.

A static disk is essentially a file containing a disk image that can be mounted on one or more virtual machines. As implied by the name, these disks are immutable and only provide read-only access to the data when mounted on virtual machines. These disks are useful for defining fixed databases for use (and reuse) for scientific calculations.

Because these are read-only files, they can be treated in the same way as virtual machine images. Consequently, the Marketplace (see below) can be used as a catalogue of available static disk images and the standard StratusLab caching mechanisms are used to minimise transfers and access latencies to these data.

StratusLab provides a separate service for persistent disks. These disks have a lifecycle—creation, use, and removal—independent of the virtual machines that use them. The persistent disk service currently sports a proprietary RESTful interface that allows control of a disk's lifecycle. The service can use either an iSCSI server (with disk storage behind) or a standard shared file system (e.g. NFS or GPFS) for the sharing of the persistent disks with the compute nodes.

The service is integrated with the compute services to allow the transparent mounting and dismounting of these disks on running virtual machines, either at boot time or while a machine is running.

The project has put a low priority on file-based storage services for two reasons: (a) most users already have access to file-based storage through grid storage services, WEBDAV-enabled servers, or similar services; and (b) a disk-based abstraction makes more sense when working at the Infrastructure-as-a-Service level that StratusLab targets.

SNIA (http://www.snia.org/) has proposed Cloud Data Management Interface (CMDI) (http://cdmi.sniacloud.com/) as a standard for storage services. This is a RESTful API for storage that has a largely file-based abstraction for storage. Treating the virtual disks as files within this API makes sense. Consequently, the persistent disk service will be enhanced to provide a CDMI interface.

6.3 Network

To connect users with distributed resources, cloud computing critically relies on a fast and stable network infrastructure. For the European Research Area, GÉANT (http://www.geant.net/) provides this infrastructure in partnership with national organisations. Commercial entities, however, cannot use the research network and must instead rely on commercial networking providers.

Advanced networking services, including dynamic Virtual Local Area Networks (VLAN), are not needed for basic use of a cloud (and in fact are not needed at all for private cloud deployments). However, for public cloud deployments, such services can increase the security and help alleviate data protection concerns.

The current releases of the StratusLab Cloud Distribution provide a very basic level of networking services, relying on the existing production network and permitting the allocation of "public," "local," or "private" IPv4 addresses. These three types of addresses correspond to three distinct use cases for the StratusLab cloud.

Public IP addresses are visible from inside and outside of a cloud deployment. They permit services running on a cloud to function as real services with visibility to users outside of the organisation running the cloud infrastructure. As they are public, they also require a higher level of security at the level of the running virtual machine. A disadvantage of public IPv4 addresses is that they are becoming increasingly scarce.

Local IP addresses are visible from virtual machines running within a cloud deployment. Access to external services from the virtual machines requires going through a NAT (Network Address Translation) server. Despite the more limited visibility of these addresses, they can be used to communicate between related machines within a cloud. Examples of this include nodes involved in parallel calculations (for example MPI calculations) and worker nodes in a batch system.

Private IP addresses are visible only from the physical host on which a virtual machine is running. Access to external services takes place through NAT on the physical host. The very limited visibility of these addresses increases the security of the machine. Despite the limited connectivity, machines running with private IP addresses can participate in master/worker frameworks such as BOINC (http://boinc.berkeley.edu/) or XtremWeb (http://www.xtremweb.net/).

Having three different types of IP addresses available allows users to balance the requirements of services running within their virtual machines with potential security improvements due to limited visibility of the services. This also allows scarce public IPv4 addresses to be conserved for those virtual machines that really need to be externally visible.

The above scheme has turned out to be very flexible, permitting a large variety of services to be deployed on a StratusLab cloud. Nonetheless, the distribution will incorporate improvements to further improve the security of running machines: provision of VLANs and dynamic firewalling. Another important topic for the future will be the support of IPv6 addressing, both for the cloud's physical hosts and for the virtual machines running within the cloud.

6.4 Marketplace

Creating robust and secure virtual machine images requires significant experience. For many people, the necessity to create a virtual machine image is a significant barrier to the use of cloud resources. Sharing virtual machine images can allow many people to overcome this barrier.

StratusLab provides a "Marketplace" that allows: (a) users to browse for relevant, pre-existing virtual machine images; (b) system administrators to evaluate virtual machine images against their security policies; and (c) machine creators to publish their work to a larger audience (see fig. 10.3 below). The Marketplace is a searchable database containing metadata entries for (both disk and machine) images; searches can be done programmatically or via a web browser. The image content itself is hosted in cloud storage or on an external server.

User Workflow: Users search for images based on the metadata, then use the image identifier to launch a virtual machine or to attach a static disk to a machine. The metadata contains basic information such as the operating system, version, and architecture, as well as service information such as what applications are available in the image. The required entries are kept to a bare minimum, instead allowing communities to adapt and to extend the metadata as necessary.

Fig. 10.3. Marketplace Workflows

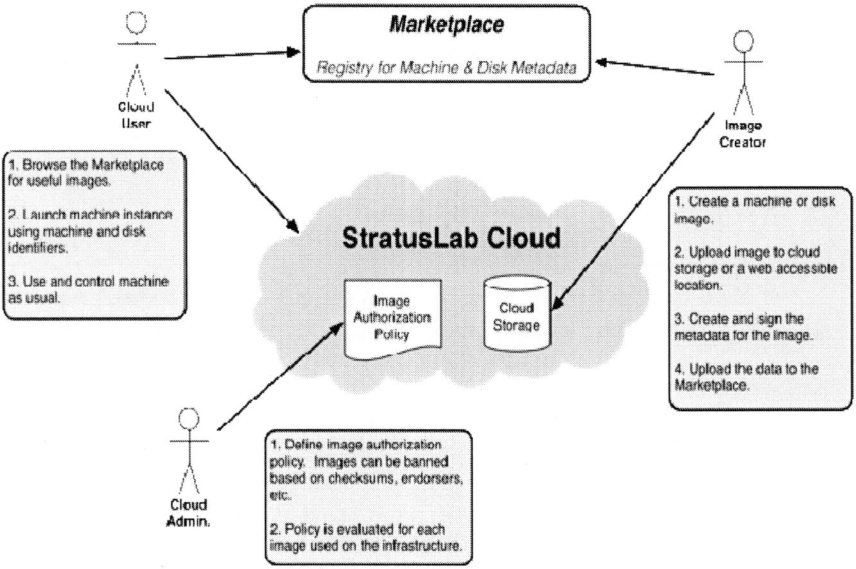

System Administrator Workflow: System administrators define a policy that validates images according to the site's requirements. All images used are then validated against the policy before being deployed within the cloud. The metadata information is cryptographically signed, allowing the administrator to decide if an image can be trusted based on the endorser and the metadata. Service information such as required ports allow the system to determine if the machine can run correctly within the given cloud deployment.

Creator Workflow: Image creators upload their image to cloud storage or make it available via an external service. They then create a metadata entry describing the image, sign the entry, and upload it to the Marketplace. In so doing, they endorse the image. This endorsement can be revoked by uploading another metadata entry that marks the image as deprecated. An interesting point is that people other than the creator can endorse the image, for example marking it as certified by an independent evaluator.

The implementation is very open, allowing the largest number of people to make use of the Marketplace. At the same time, it permits the uploaded information to be validated and evaluated against security criteria. See the Marketplace Technical Note (StratusLab 2011) for the complete list of security concerns.

The core functionality of the Marketplace is in place. Changes are likely to come in the content of the metadata entries as the user community discovers what information is necessary to effectively share images. It is also likely that social features will be added to the Marketplace, allowing the reliability of images and endorsers to be determined by the community.

6.5 Claudia

Using an IaaS cloud service is effective for single machine services and for building platforms with multiple cooperating machines.

For platforms however, it often becomes tedious to have to synchronise the deployment of the various machines (for example, servers must be deployed before clients) and to provide manual failover strategies. High-availability and auto-configuring services can reduce the need for manual interventions. Unfortunately, such services are still very much in the minority at most data centers.

Claudia, a "service manager" aims to automate the deployment of an ensemble of machines (a "service") and to provide limited autoscaling of the service. It uses an OVF (DMTF 2009) description that describes both the dependencies between machines and rules for changing the deployment based on performance or load metrics.

Claudia is a part of the StratusLab production release and has been used to automatically deploy a batch system and a complete grid site. It currently does not use the standard StratusLab authentication mechanism and is thus appropriate only for private cloud deployments.

The service exposes the TCloud (Telefonica 2010) interface, which has been submitted to the DMTF (http://dmtf.org/) as a proposed standard. As the service evolves, the standard authentication mechanisms will be incorporated to allow use on public cloud deployments. Incremental improvements in the functionality and robustness of the code will continue over the lifetime of the project.

7. Reference Infrastructure

Although the project does not intend to provide a perennial e-infrastructure for scientists, it does provide a reference cloud infrastructure to demonstrate the StratusLab cloud distribution, to allow production use of the software, and to encourage feedback on the project's software.

This infrastructure is provided by GRNET in Athens, Greece, consisting of 17 physical servers, 256 CPU cores, 768 GB RAM (48 GB/node), 3.6 TB of disk space, and 10 Gb/s network interconnects. To ensure feedback from a wide variety of people, the infrastructure is open to the public (with account registration).

A second infrastructure will also be available at LAL in Orsay, France to give users access to a second infrastructure and to allow the project to investigate various cloud federation schemes.

8. Use of the StratusLab Distribution

The production release is recent, and real world deployments and use of the StratusLab distribution are just beginning. Nonetheless, there are indications that the cloud services are robust and meet the needs of users.

The operation of the reference infrastructure is the first indication. To demonstrate further the functionality and stability of the StratusLab distribution, GRNET also runs a production grid site (HG-07-StratusLab) within the European Grid Infrastructure on top of the reference cloud infrastructure. This site had to go through all of the standard certification procedures to be a part of EGI and is monitored through its standard operational tools. This site validates the model of deploying grid services as a PaaS over a StratusLab cloud and the functionality of the distribution itself. The services have remained stable for more than six months, even in the face of frequent upgrades of the StratusLab software.

A few bioinformatics researchers and engineers act as a pilot scientific community for the project. They have created customised virtual machine images that allow them to access standard bioinformatics databases and to analyse this information with standard software tools within the domain. A more involved analysis, involving the determination of protein structures based on Nuclear Magnetic Resonance data, will determine if the software can effectively handle complicated workflows and permit easy autoscaling in response to the changing needs of the analysis.

SixSq is also currently installing a StratusLab cloud instance to test whether it can run the standard software environment used by the European Space Agency. The ultimate aim is to use cloud technologies to expand the scope of the testing for their software and to speed the deployment of new releases.

Feedback from all of these installations and tests will undoubtedly improve future releases of the distribution and lead to more deployments of the software.

9. Challenges

The current StratusLab distribution provides a solid, production-quality cloud distribution. Nonetheless, there are some challenges to running a cloud that will require further software development and changes in the way non-commercial computing resources are managed. Some aspects of these challenges are being explored within the project, but full solutions will need research and developments extending beyond the end of the project.

9.1 Scheduling

Although the usual cloud usage model does not expose queues or scheduling to the end-user, scheduling is critical for cloud providers. On cloud infrastructures, two areas require scheduling decisions: initial placement of virtual machines and optimisation of the total infrastructure.

The initial placement of a virtual machine is very similar to scheduling a job in a traditional batch system. The scheduler must match the requirements of the virtual machine against the (available) capacity of physical machines making up the cloud. All of the accumulated knowledge and experience with batch systems is directly applicable to creating an efficient cloud "placement service."

However, virtualisation technologies also permit virtual machines to migrate from one physical host to another. This opens up new possibilities for adjusting the load on a cloud based on any number of characteristics: energy use, service level agreement management, state of physical hardware, etc. However, migration itself has resource and service availability overheads that must be considered when balancing the load of the system. Developing efficient "optimisation services" for the cloud are critical for efficient management of the resources and driving down the costs of providing them.

9.2 Infinite Capacity

Simplicity and elasticity are the two fundamental characteristics of cloud computing; without them the technology is much less compelling. Both come from a model in which the cloud resource has infinite capacity. In this case, resource requests can be handled immediately, no matter how numerous or large. This simplifies the usage model, avoiding complications like queuing, scheduling, and priorities that are features of other shared computing resources.

No real cloud truly has infinite resources, but for all intents and purposes, large public cloud providers like Amazon or Microsoft have enough to make them appear infinite. Resource centres typical of universities or small companies, however, do not. If these centres are to deploy their own clouds and give their users the same simple usage model, they must maintain the illusion of infinite capacity. How can this be done?

Federation of cloud resources offers one possible solution (see fig. 10.4 below). One model has different cloud administrators agreeing to share resources. In this case, requests which cannot be handled by one cloud resource would be outsourced to another peer cloud. This would be completely transparent to the user and maintain the illusion of infinite capacity. This model is usually called "cloud bursting" when commercial public providers are used for the outsourcing.

Another model for federation would expose the individual cloud resources to the user. In this case, if one cloud was at capacity, then the user could choose another cloud resource. No doubt, cloud-brokering services that search for and select particular clouds would quickly spring up, hiding this complexity from the users. This model is very similar to the "brokering" model that is used in current large-scale grid infrastructures, like EGI.

Federation, however, brings its own issues. Many virtual machines work in concert with others or are co-located with the data they process. When outsourcing virtual machines to another cloud, these interdependencies are crucial as network bandwidth is much lower over the wide-area network than within a single cloud resource. In addition, the shipping of data between sites may compromise privacy or confidentiality constraints. Technical solutions to these issues can be found, but these solutions will significantly complicate any federated cloud deployment compared to a single site cloud deployment.

Reintroducing queuing to the cloud usage model offers another partial solution. Not all resource requests need to be handled immediately. Those less urgent tasks could be queued until demand drops and resources become available. The tricks are (a) to identify those tasks that can be queued and (b) to keep the standard (non-queued) usage model simple. Amazon provides a rather elegant solution with its spot instances—virtual machines that run only when the price drops below a certain threshold.

Fig. 10.4. Cloud Federation Models

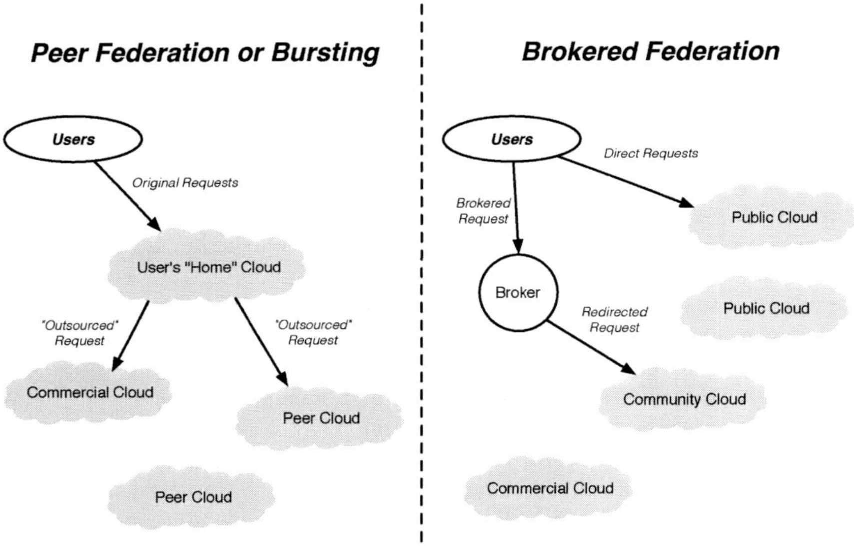

9.3 "Pay as You Go"

All public, commercial clouds have accounting and billing services that allow users to pay for only the resources that they consume. This is a core part of their business model and source of their revenue. This economic model also ensures that users monitor their utilisation and gives the provider control over the total utilisation.

Because cloud providers identify their users through their credit cards and tie utilisation directly to the cards, users are highly motivated to monitor their utilisation, using the most efficient methods for calculations and shutting down unneeded virtual machines. This encourages efficient sharing of the capacity.

Within the economic model, providers have control over the prices. This gives those providers the ability to control the total load of their resources by raising or lowering prices. Essentially this simplifies the load management both for the provider and for the users, by avoiding complicated mechanisms using quotas and fair shares.

Economic models can offer the same benefits in academic settings. The problem is not technological—all public cloud providers deploy and use accounting and billing systems. The problem is overcoming the reluctance to using these models and the inexperience in pricing resources in academic environments. Nonetheless, moving to such a utilisation model (even with "fake" money) would facilitate equitable sharing within an institute and federation between cooperating institutes.

10. Future Directions

Within the first year, the project created a complete, open source cloud distribution, providing the computing, storage, and networking services required for deploying an IaaS cloud. The distribution also includes high-level services like the Marketplace to promote sharing of machine images and Claudia to orchestrate the deployment and control of multiple machine services.

The second year will bring incremental changes to the existing functionality, focusing on improving the ease of use and integration between the services. The services will evolve to support the emerging cloud standards: OCCI, TCloud, CDMI, and OVF. This transition will be transparent for users of the StratusLab command line interfaces.

More profound functional additions to the StratusLab distribution will come from the support for federated cloud infrastructures. This will include cloud "bursting" to commercial public cloud providers or to other StratusLab cloud deployments. It will also include a study on how best to include and to federate cloud resources on the European Grid Infrastructure.

Functional advances are of little benefit without real users of the software. Consequently, a major focus of the second year of the project will be to identify concrete scientific and engineering use cases and to ensure that they can be satisfied using a StratusLab cloud. Already, scientific use cases have been identified in bioinformatics and computational chemistry. An engineering use case involves the deployment and use of the typical software development tools (code versioning system, bug tracker, web site, wiki, LDAP server, etc.) to create a software engineering PaaS.

Longer-range plans include measuring the performance of specific applications on a StratusLab cloud and identifying areas that need improvement. This would allow cloud services to offer High Performance

Computing services and expand the range of applications that are well adapted to running in a cloud environment.

11. Summary

Cloud technologies provide many benefits for scientific and engineering applications: customised execution environments, near-instantaneous provisioning, elasticity, and the ability to run user-level services. To provide elasticity while remaining efficient, there is a push towards larger, centralised data centers. However, there are also many barriers that prevent a rapid shift back to centralised computing. On the one hand are the current capital investments and human investments in existing data centers in research institutes and university departments. On the other are concerns about data management, especially the production, availability and maintenance of large scientific data stores, in a centralised model when much of the data is produced in a distributed manner. Because of these barriers, the shift towards a more centralised model based on public cloud infrastructures will be gradual.

If scientists and engineers are to take advantage of the benefits of cloud computing in the short term, their institutes must be able to transform their current resources into a cloud infrastructure. This requires a complete, readily accessible cloud distribution that they can install.

StratusLab has developed and maintains a complete, open source cloud distribution that allows people to install their own public or private IaaS cloud infrastructure. The StratusLab services include the computing, storage, and networking services required for an Infrastructure as a Service (IaaS) cloud. It also includes services like the Marketplace that facilitate sharing of machine images and Claudia that allows the deployment and management of complete software systems.

The project produced four beta releases of the software and its 1.0 production release in the first year of the project. This was possible because it chose to integrate existing solutions where possible (for instance using OpenNebula for the virtual machine management) and to follow an agile software development method that permitted rapid evolution and validation of the StratusLab distribution.

Deploying a cloud infrastructure in the smaller data centers of research institutes and universities does pose some general (not entirely technical) challenges. Elasticity is a key characteristic of cloud computing, but how is this maintained when deployed on finite resources? Federation is a possible answer here, but this can bring complications both for the cloud implementers and users. Virtual machine scheduling policies are extremely important to meet certain goals, like availability, performance, and energy efficiency. How can these be incorporated into cloud deployments? Most commercial providers use a "pay as you go" economic model, both as a means of monetising their infrastructure and as a way to control resource occupancy. How can this be adapted to academic environments that do not have a tendency to use such techniques?

The StratusLab cloud distribution will continue to evolve, incrementally adding new features and improving existing ones. With the initial 1.0 production release as a firm base to build upon, the project will concentrate on supporting hybrid cloud infrastructures and on demonstrating real scientific and engineering use cases. Many of these use cases will eventually become domain-specific platforms running on a StratusLab IaaS cloud infrastructure, allowing European e-Infrastructures to become more diverse while maintaining a common infrastructure to promote sharing and collaboration.

Acknowledgements

StratusLab is co-funded by the European Commission through the 7th Framework Programme (Capacities), contract number INFSO-RI-261552. The authors also acknowledge the support provided by other members of the StratusLab collaboration and by the institutes involved in the project.

References

Svensson, Peter. "What Amazon's Outage Means for Cloud Storage." The Huffington Post, April 21, 2011.

Mell, Peter and Grance, Timothy. "The NIST Definition of Cloud Computing (Draft)." Special Publication 800–145 (Draft), available from http://csrc.nist.gov/publications/drafts/800-145/Draft-SP-800-145_cloud-definition.pdf (accessed October 17, 2011).

Fielding, Roy Thomas. "Architectural Styles and the Design of Network-based Software Architectures." PhD Thesis, University of California, Irvine, 2000.

StratusLab. "StratusLab Marketplace", TN-Marketplace (V3.0), 2 March 2011, available from http://stratuslab.eu/lib/exe/fetch.php/documents:marketplace-v3.0.pdf (accessed October 17, 2011).

DMTF Distributed Management Task Force. "Open Virtualization Format Specification", DSP0243, 2009.

Telefónica Investigación y Desarrollo S.A. Unipersonal. "Telefónica's TCloud API Specification", Version 0.9, 26 January 2010 available from http://www.tid.es/files/doc/apis/TCloud_API_Spec_v0.9.pdf (accessed October 17, 2011).

CHAPTER ELEVEN

BUSINESS APPLICATION GOVERNANCE AND SLA MANAGEMENT IN PAAS CONTEXT

FRANCESCO D'ANDRIA, ILKNUR CHULANI
AND PHILIPP STRUBE

A cloud environment introduces a tremendous amount of flexibility and scalability with a consequence that a list of items has to be managed more closely compared to traditional systems. Platform-as-a-Service (PaaS) is a combination of a development platform and a solution stack, delivered as a service on demand. Availability, security, Quality of Service (QoS), business application management, privacy, location of cloud services and compliance are just some of the aspects of the cloud that have to be monitored and managed closely. Within the Cloud4SOA research project (http://cloud4soa.eu/), the consortium aims to face new challenges investigating new service engineering paradigms, methods and tools supporting cloud application governance in PaaS context.

1. Introduction

Platform as a Service (PaaS) is a quite new paradigm that enables independent software vendors (ISVs) or organisations to create, run, integrate and manage business applications, using services provided by third party. The actual PaaS market is in constant motion. However, it is still quite young, chaotic and highly fragmented and only a partial consolidation process is foreseen in the coming years. Gartner states, in many reports, that 2011 "is the year of PaaS," however best practices as well as standards have only just begun to be discerned.

Cloud4SOA, a European Co-Funded FP7 project, has the challenge of opening up the cloud market to small-to-medium PaaS providers in Europe, strengthening their market position, and in parallel helping alleviate the vendor lock-in barrier for cloud developers.

Cloud4SOA will address the challenge of enhancing the cloud-based application development, deployment and migration by semantically interconnecting Platform as a Service (PaaS) offerings across different cloud PaaS providers sharing the same technology, and will facilitate the access and lifecycle management for cloud-based application developers to the PaaS offering that best matches their computational needs.

In detail, the Cloud4SOA project examines semantic interoperability needs as well as business application management needs related to:

- The deployment of instances of business application in several and competitive platform offerings (Deployment Scenario)
- The federation of run-time information between the running instances in a harmonised way
- The management of business application instances
- The migration of an instance (i.e. business logic and data) between competitive platform offerings.

In this chapter we will present Cloud4SOA's approach towards a flexible and semantically interoperable application governance in PaaS context. Governance is an often-misunderstood term. Some people use it to refer to application lifecycle governance, that is, governing the lifecycle of services from creation through deployment. In the scope of the Cloud4SOA project, governance is also about delivering and migration. The methodology adopted follows an approach that integrates application governance model with Service Level Agreement Management taking into account the process of decision making, criteria and policies involved in the semantic setting-up, construction, acquisition, development, operation and management of a cloud computing capability.

We will start this chapter by presenting a market analysis of the current PaaS solutions in section 2, and the Cloud4SOA Reference Architecture in section 3. We will then focus on the Governance and SLA Management layer. Section 4 provides a background on SLA Management and WS-Agreement, while a semantic enabled approach for SLA Management is presented in Section 5. Sections 6 and 7 focus on the Cloud4SOA Governance layer requirements, use cases, and architecture. Finally, we summarise the business benefits of this research in Section 8.

2. Current PaaS solutions Analysis and Evaluation

The field of Platform as a Service providers has seen a massive growth during the past two years. The sales of Heroku to Salesforce in late 2010 in particular has given a massive boost to the market, as new entrants try to adapt the Heroku PaaS model to other languages. At the same time, larger players like Microsoft and Google have pushed forward their solutions, and IaaS companies like Amazon (Beanstalk) or Rackspace (Openstack) have tried to move up the stack with new solutions. Combined, this creates a number of different approaches to PaaS. In this section we try to give an overview on the existing PaaS solutions and compare the players according to the evaluation criteria selected. We limit our selection to public cloud providers, i.e. PaaS solutions that are readily set up and can be used immediately, and do not include solutions that create a private PaaS environment.

2.1 Evaluation criteria

The list of evaluation criteria for the PaaS solutions considered as relevant for the Cloud4SOA project are:

- **Life cycle stages**. Applications have different life cycles in the range from developing-testing-operating-retired-deleted.
- **Vendor Lock-in grade**. The use of open specification and technologies is a key point in switching between different PaaS providers. Lock-ins prominent at one provider might hinder interoperability between one service and another and might require adaption work.
- **Application governance capabilities**. Application governance describes the ways in which application stakeholders can access or edit the application or underlying account settings. Typically, PaaS providers often offer two kinds of governance interfaces: a command line interface (CLI) for the developer and a graphical user interface (GUI) for administrating account specific items.
- **SLA management**. SLAs play a vital role in determining the suitability of a service
- **Real time metrics available**. The metrics available for monitoring the applications performance on a real time basis
- **Historical reports available.** The metrics available for monitoring on a historical basis

- **Billing methodology**. Describes how services are paid for
- **Languages supported**. Describes the languages that are supported by the platform
- **Additional platform services**. Almost all web applications need additional services besides raw computing power, for example databases. PaaS providers need to offer those in order to make interoperability possible.
- **Geographical locations**. The geographical location of the platform is an important aspect when it comes down to latency issues or personal data security regulations
- **Code upload**. The code upload component is the key interaction point for developers working with the platform. By adhering to the same code upload style and tools (for example version control systems like Git), the interoperability from the developer standpoint can be guaranteed
- **File System available**. The availability of a file system is often required by modern applications.
- **Single tenant/multi-tenant**. Single tenant platforms can allow customers to specify or modify underlying hardware architecture. With multi-tenancy platforms this is not possible since the abstraction level is on the application layer. For security reasons, modifying the underlying hardware configuration or allowing certain types of setting are excluded in an multi-tenancy environment.
- **Auto scaling**. PaaS providers can offer to automatically provide more resources as the load on an application increases.
- **Underlying hardware environment/IaaS provider**. Many PaaS providers build on top of popular IaaS providers like Amazon AWS. This creates an eco-system of providers enabling users to mash up services, for example use a PaaS provider and a privately operated service in the same hardware environment, because problems like latency and regional difference are non-existent.

2.2 Evaluation Conclusion

Due to the short time frame of the existence and evolution of PaaS solutions, and the various backgrounds of the different providers, the market offerings show a fairly diversified picture on how the PaaS solutions are operated and administrated. A unique set of standards has not yet crystallised. This remains a weakness of the market, making it difficult for end customers to switch from one provider to another, unless manual labour is invested. However, some trends clearly seem to emerge. For example, the lack of a vendor lock-in is widely adopted by many platforms, enabling users to (at least theoretically) port their application between different PaaS solutions. Likewise, interacting with the platform is often done mainly with the help of a command line interface (CLI), though the commands differ from platform to platform. In addition, the use of a version control system to administer the code has been widely adopted, whereby Git tends to be the most widely used system.

Most providers have implemented an add-on approach, i.e. providing databases or other additional services in the form of differently sized packages that can be added to the application. In terms of programming languages, most providers currently focus on one language but many have announced the support for more languages in the futures (e.g. PHPFog, cloudControl, Heroku, Dotcloud). The biggest differences amongst hosted PaaS solutions exist on the billing and resource usage end. Most providers have a unique method of billing consumption and make it hard to compare. Likewise, the historical data available and the format how this data can be received differ significantly between providers.

3. Cloud4SOA Reference Architecture

A key outcome of the Cloud4SOA project is the design and implementation of a Reference Architecture that aims to resolve semantic incompatibilities raised between heterogeneous cloud computing platforms. The Cloud4SOA reference architecture is based on open standards and:

- Semantically interconnects heterogeneous PaaS platforms
- Provides intelligent and adaptive front-ends to connect users to the management layer
- Provides capabilities to publish and discover PaaS offerings as well as helps users to deploy, execute and monitor business applications on them (through the virtualization layer)
- Allows the governance of business application lifecycle and helps manage Service Level Agreements

Fig. 11.1. Abstract view of Cloud4SOA Reference Architecture

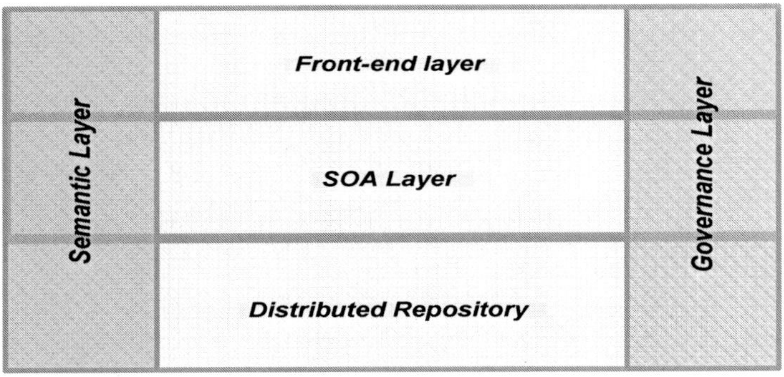

To this end, it utilises the knowledge derived from the generic architecture of cloud computing PaaS systems. Specifically, it should capitalise on IT trends, such as SOA, semantics and intelligent frond-ends. The envisioned Cloud4SOA Reference Architecture (fig. 11.1 above) comprises of three horizontal layers, namely the Front-end Layer, the SOA Layer, the Distributed Repository, and two vertical layers, namely the Semantic Layer and the Governance Layer, that span across all the horizontal layers.

- The Front-end Layer supports the user-centric focus of Cloud4SOA and provides Cloud-based application developers easy access to the Cloud4SOA functionalities;
- The SOA Layer implements the core functionalities offered by the Cloud4SOA system, such as service discovery, announcement and deployment;
- The Distributed Repository puts in place the technical infrastructure, for example distributed virtualised registries, on top of which the Cloud4SOA system is built;
- The Semantic Layer provides the semantic models and tools offered by Cloud4SOA; and
- The Governance Layer offers a toolkit for monitoring the service life-cycle of Cloud4SOA services.

While an abstract view of Cloud4SOA Reference Architecture is shown in fig. 11.1, fig. 11.2 below shows a refined view of the architecture.

Fig. 11.2. Cloud4SOA Component Diagram

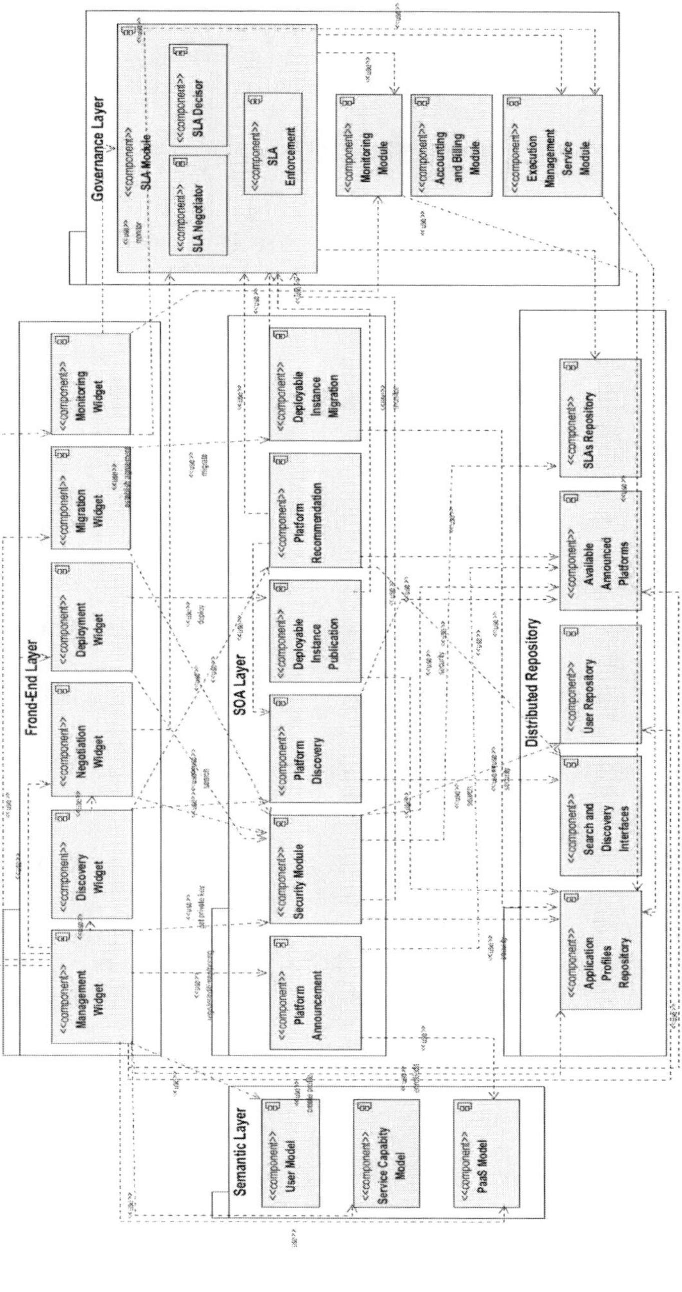

1.1 Cloud4SOA Physical Architecture

The Cloud4SOA physical architecture comprises of two basic physical "entities": (a) a centralised system; and (b) a (distributed) set of "**Platform Adapters**."

The centralised system is practically an n-tier web application which contains:

(1) A web 2.0 User Interface layer that facilitates the interaction of Developers and PaaS owners with the system. Interaction refers to adding/editing platform/application profiles, initially deploying archives, performing deployment queries, govern applications, visualising monitoring data, etc.

(2) A Business Logic layer (a.k.a. SOA layer because it is exposed through Web Services) responsible for undertaking/implementing PaaS query handling (i.e. matchmaking), new PaaS announcement, publication of new Platform/Application models, transformation of a submitted application in order to be deployed/migrated at a chosen PaaS, federation of monitoring data from distributed PaaS platforms, etc.

(3) A Persistency layer that comprises of two physical subcomponents;

   (a) One Database/Triple Store that stores Business Logic-related data such as user-accounting and semantic information such as semantic models etc.

   (b) One queue that is used to push asynchronously information to a Specific platform Adapter

The "Platform Adapter" is the Proxy between the Cloud4SOA central system and the various PaaS platforms. The API that a "Platform Adapter" exposes to the central system is unique (i.e. the one that Cloud4SOA consortium has defined). However, this API can sometimes be satisfied by the low-level PaaS-specific API (i.e. only a "wiring" is necessary between the specific API to the Cloud4SOA one) and sometimes not. In case of incompatibility, either the "missing part" is hardcoded by the PaaS provider (or developer), or is intentionally left unimplemented.

From the functional perspective Adapters are used to:

(1) Push Asynchronously Information, i.e. Deploy/Un-deploy archives to a PaaS platform

(2) Push Synchronously Information, i.e. stop/start application instances for maintenance purpose

(3) Pull Asynchronously Information, i.e. migrate applications to another PaaS provider

(4) Pull Synchronously Information for monitoring and SLA enforcement purpose, i.e. get status of application_1 that is installed at PaaS_platform_1

Consequently, both synchronous and asynchronous communication patterns are used for the communication between the adapters and the central system. Therefore a database and a queue are necessary. Platform Adapters piggyback on deployed applications and store monitoring information locally. This information is periodically pushed to the central system. Finally, platform adapters constitute an overlay network that is used only during migration in order to avoid transfer through the central system.

4. SLA Management Background and State-of-the Art

The management of Service Level Agreements (SLAs) is a key aspect in distributed resource management. It has been the focus of many research and development projects, either industry-driven projects or as part of national as well as international research activities. In the scope of Cloud4SOA project, SLAs represent a contractual relationship between independent software vendors (ISVs), consuming virtualised Platforms, and the PaaS providers. SLAs describe the resource that is delivered, the functional and non-functional properties of the resource, and the obligations of each party involved (for example which party is the provider of the service). Additionally, SLAs define guarantees for the functional and non-functional resources properties. These guarantees specify a service level objective that must be met in order to fulfill a guarantee, and they define compensations in form of penalties and rewards which are charged in case the guarantee is fulfilled or violated respectively.

The Web Service Agreement (WS-Agreement) standard of the Open Grid Forum (OGF) (http://forge.gridforum.org/projects/graap-wg) is a novel and well-accepted standard for creating and enforcing SLAs in distributed environments as well as monitoring their resources properties. An interesting report summarises the state of SLA work in past European projects (CoreGrid 2007). In that report, the authors compare the current usage of SLA (including in business) in six European commission's FP6, where the issue of extending WS-agreement for negotiation/renegotiation is highlighted. The Akogrimo project (http://www.akogrimo.org/) (D'Adria 2006) and BEinGRID project (http://www.beingrid.eu/) (D'Adria 2008), or BREIN project (http://www.eu-brein.com/) have successfully shown the integration of SLA management subsystems with similar characteristics, but addressing different viewpoints of SLAs that could be seen as being complementary.

The WS-Agreement implementation for Java (WSAG4J) is an implementation of the WS-Agreement protocol defined by the OGF (https://packcs-e0.scai.fraunhofer.de/wsag4j/). It provides comprehensive support for common SLA management tasks such as SLA template management, SLA negotiation and creation, and SLA monitoring and accounting.

4.1 WS-Agreement

The paper (Kübert et al. 2011) gives an exhaustive summary of recommendation. As previously mentioned, the WS-Agreement standard is a full recommendation of the Open Grid Forum (OGF) defined by the Grid Resource Allocation Agreement Protocol working group (GRAAP-WG). The OGF is a community "committed to driving the rapid evolution and adoption of applied distributed computing" (http://www.ogf.org/). It has published more than 150 documents on, for example, the well-known Open Grid Service Architecture (OGSA) specification, the Open Cloud Computing Interface (OCCI) specification, grid extensions to the FTP protocol (GridFTP) and many more (http://www.ogf.org/gf/docs/?final\&all).

The objective of the WS-Agreement specification is to define a language and a protocol for advertising the capabilities of service providers, creating agreements based on templates, and for monitoring agreement compliance at runtime. An agreement between a service consumer and a service provider specifies one or more Service Level Objectives (SLOs) both as expressions of requirements of the service consumer and assurances by the service provider on the availability of resources and/or on service qualities. An agreement life cycle includes the creation, monitoring and termination of agreement. WS-Agreement describes an XML schema for specifying service level agreements, for example between a service provider and a consumer, as well as a simple protocol that describes how parties can communicate in order to build an agreement. WS-Agreement builds on other web service specifications, namely the following:

- Web Services Addressing (WS-Addressing) 1.0 Core
- Web Services Resource Properties (WSRF RP) 1.2
- Web Services Resource Lifetime 1.2 (WSRF RLF)
- Web Services Base Faults 1.2 (WSRF BF)

Elements of these specifications are included into WS-Agreement, for example the Endpoint Reference from WS-Addressing which describes how resources can be addressed, or fault types from Base Faults which define error messages that can be thrown by operations. Other parts of these specifications are used to extend the interface that WS-Agreement specifies by itself. This is done by including port types described by WSRF RLF and WSRF RP, which can be thought of as inheriting from these specifications as super classes. WSRF RLF provides operation for specifying the life time of a resource while WSRF RP describes standardised operations to get, set or query so-called "resource properties"; they can be thought of as getters and setters in Java, which are instance methods that retrieve or store values from object instances.

5. Semantic-enabled approach for SLA management

The current WS-Agreement specification is based on XML based domain vocabularies and therefore limits the ability of matching the agreements to syntactical matching. Pure XML-based representation of information does not enable machines to access the intended semantics of XML messages. Thus, semantics is still only comprehensible by humans (Krummenacher et al. 2005). As a result, systems are rigid and cannot be adapted to changes without human intervention (Fensel et al. 2006), as reflected in activities regarding services like service discovery, invocation and composition.

The Cloud4SOA approach proposes using domain knowledge captured using ontologies and rules to extend the matching capabilities beyond simple string matching. A matcher considering only the syntax of the agreements without the domain knowledge may not able to correctly identify all matches. We illustrate the usefulness of our approach with the following example. PaaS is usually scaled and billed through parallel copies of the application running (maybe called boxes, dynos, processes, etc). Consequently, it makes a huge difference in terms of availability if multiple copies are on the same or different servers. Another key point to take into account is that the PaaS providers automate a lot of steps for platform consumers, for example deployments. Consequently, it makes a huge difference whether each deployment takes ten seconds or ten minutes.

Consider that an application provider (PaaS consumer) has the following requirements:

- two parallel copies of the application running on the same server
- application deployment is lower than 30 seconds

and a PaaS provider offers the assurances of:

- "X" boxes free on a same server (where "X" is higher than 2)
- That application deployment time is proportionately with its size.

A syntactic matcher would perform a string comparison to determine if the PaaS provider can satisfy the platform consumer's request. A syntactic matcher would generally determine that these two requirements do not match on the grounds that the PaaS provider does not provide an assurance for availability. However, the Cloud4SOA approach would utilise an ontology which provides a deeper understanding of the domain with the help of domain rules.

Moreover, the terminology to define the requirements and offers of services in SLA can vary between the customers and providers. For instance, business customers try to maximise their goals requiring business terms like high availability, high performance, low price while PaaS providers need to define and subsequently map their offers in terms of the IT metrics about the infrastructure they will use (storage capacity, hardware resource properties). Thus, the previous heterogeneity problem is increased, considering the different abstraction layers of the entities involved.

Taking into account all these factors, further requirements have arisen:

- Address interoperability issues (vocabulary domain, national languages, etc) between PaaS providers and PaaS consumer in order to overcome the problem of different terminologies.
- The proposed solution should be compatible with the selected specifications (WS-Agreement).

5.1 Proposed solution

During the preparation of this document, outcomes of several projects and the approaches they followed were still being analysed and evaluated.

Some of them, such as BREIN (http://www.eu-brein.com/), proposed a flexible way to negotiate service level agreements using a semantic annotation (www.ontotext.com/kim/semantic-annotation) for SLA specification. Mainly, this approach involves the annotation of SLA templates by using SLA ontology concepts as well as the new SLA terms created by providers. Concretely, the annotation methodology is based on the lightweight annotation mechanism adopted in SAWSDL. The Semantic Annotations for WSDL and XML Schema (SAWSDL) W3C Recommendation defines mechanisms using which semantic annotations can be added to WSDL components. SAWSDL does not specify a language for representing the semantic models, for example ontologies. Instead, it provides mechanisms by which concepts from the semantic models are defined either within or outside the WSDL document and can be referenced from within WSDL components as annotations. These semantics, when expressed in formal languages, can help disambiguate the description of Web services during automatic discovery and composition of the Web services.

SAWSDL is a simple extension of WSDL using the extensibility elements. The semantic annotation technique practically consists of keeping the original WS-Agreement schema almost the same. It only adds some additional attributes, which provide references to semantic models that describe the meaning of the related XML content. In the case of structured data, identified parts can be related to ontology classes or other semantic constructs. In the case of XML data, these relations can be provided as additional XML attributes (Jacek et al. 2007). The diagram below outlines the basic components needed for the implementation of the pattern:

- The Editor makes it possible for users to add or edit annotations to the XML data,
- The Parser for XML data parses the data and makes the content interpretable and usable for other components
- The Parser for Semantic Annotations helps the XML parser to provide enriched interpretation facilities for the XML data.

Fig. 11.3. Semantic annotation

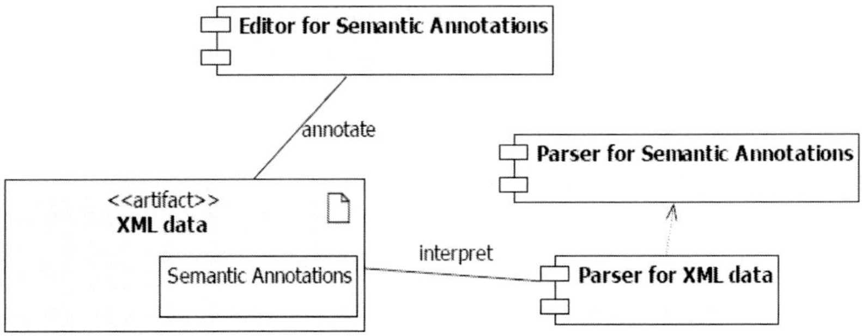

A further interesting approach would be to work on SLA and Quality of Service ontologies and rules applicable to the PaaS context, to match PaaS consumer and PaaS offering requirements in the contract negotiation phase. These ontologies would help the Cloud4SOA system to offer a description much closer to human (natural) language than machine language based specifications, like WS-Agreement.

A second application topic is at runtime during the QoS enforcement and detection of SLA violations. In that case, measurements are made to detect the real working behaviour of the platform and when a violation happens, it is translated to the platform consumer.

Since, our analysis and evaluation is not finished yet, the final approach will be described in the next phases of the project.

6. Cloud4SOA Governance layer

The focus of the Cloud4SOA Governance Layer is to define, design and implement the Service Lifecycle Governance Framework, part of the more generic Cloud4SOA Reference Architecture. In the scope of the Cloud4SOA project, the Governance layer is responsible for implementing Business Application Governance (BAG) in platform environments. Although, in the context of the Cloud Computing Governance, it represents a new paradigm, in many respects it extends the SOA governance paradigm. Both cloud computing and SOA are predominantly accessed using Web service APIs, have Service-Oriented architecture at their core and connect consumers and providers through a service Level Agreement (SLA). All of the above follow a lifecycle based approach for executing business application governance in PaaS, described below and implemented during the project lifecycle.

Fig. 11.4. Business Application Governance Lifecycle

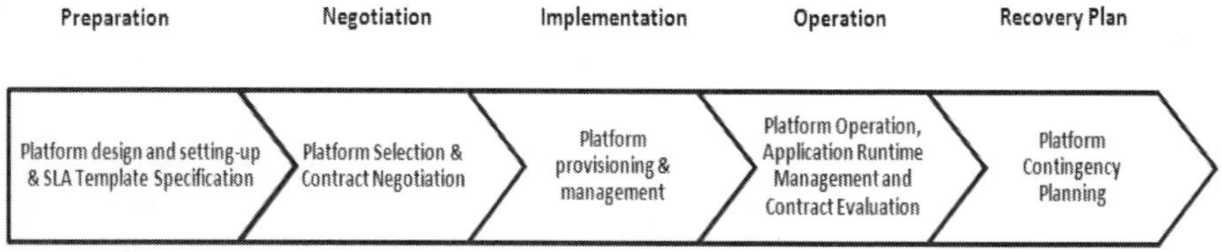

As seen in figure 11.4 above, Governance PaaS lifecycle can be described in five main processes:

- **Platform design and setting-up & SLA Template Specification**: Describes the processes of platform preparation, including publishing service advertisements and storing service-related artefacts. For the resource provider, a clear step-by-step procedure describing how to write an SLA template to provide with correct (and possible legal) resource offering description.
- **Platform Selection & Contract Negotiation**: Describes the processes of platform as a service comparison and selection, contract negotiation (among platform providers and platform consumers) which must include the SLA definition, Quality of Service (QoS) definition, and appropriate security, business assurance and operational requirements.
- **Platform provisioning & management**: Describes the processes of creating and managing policies and associating them with application as well as installation and readiness, provisioning and management.
- **Platform Operation, Application Runtime Management and Contract Evaluation**: Processes and policy focuses on the operational management of the business application running on the platform. It takes into account resource metering and monitoring, the SLA contract (associated to the Business Application) enforcement, and all related operational and management processes including accounting and billing procedure based on business models.
- **Platform Contingency Planning**: A mechanism for alerting and reacting to abnormal situations so that SLA Management can undertake corrective decisions according to defined policies. It focuses on the governance processes and activities that enable a

robust, reliable and agile environment, i.e. application migration to a different cloud PaaS Provider.

6.1 Requirements of the Governance layer

The methodology followed for the design of Cloud4SOA governance layer architecture is based on the identification of the key requirements the architecture needs to address. Subsequently, the requirements have been translated into use cases and finally in system functionalities and components.

This section documents the functional requirements per stakeholder. In the context of this chapter, only the top and medium priority requirements are listed in table 11.1 below.

6.3 Use Cases analysis

A total of six use cases are described in this section and are discussed in table 11.2 below. Tom Pérez is the actor in these use cases. Tom is a freelance software developer who has produced a new solution for Financial Risk Analysis as a plug-in to a well-known package.

Table 11.1. Cloud4SOA Governance layer requirements

Stakeholder	Requirement	Priority
Cloud-based appli-cation developer	Search for available PaaS offerings using specific criteria	Top
	Monitor, manage and configure a cloud-based application's lifecycle (deployment, federation and migration)	
	View details on the infrastructure and the services provided by each Cloud4SOA-enabled platform	
	Be able to migrate an application from one Cloud4SOA-enabled platform to another	
	Receive notification when an SLA violation is raised	
	Establish agreements with Cloud4SOA enabled providers (when registered)	Medium
	Deploy (part of) application on a Cloud4SOA enabled platform	
	Rate PaaS providers and the service offered by them (in terms of quality, reliability etc.)	
	View the user-generated comments and ratings for a specific PaaS provider	
Cloud PaaS provider	Define generic pricing model and the associated performance characteristics (semantic profile of the platform) and publish them	Top
	Establish agreements/form alliances/collaborate with other Cloud4SOA enabled providers	
	Be rated negatively when violating an SLA Agreement	

Table 11.2. Cloud4SOA Governance layer use cases

No	Use Case Name	Short Description
1	**PaaS Search and Negotiation**	Tom wants, first of all, to install and test the solution in a PaaS environment. Since he already has a valid profile in the Cloud4SOA community, he logs-in to the Cloud4SOA portal, and through the Search Widget, he makes a platform search, providing: • Application information, • Service Level Agreement needs
2	**Application Governance**	(1) **Deploy**: Tom wants to install and test the solution in a PaaS environment. He has already logged-in to the Cloud4SOA system, performed a search to find the PaaS providers that matches his application's needs, and selected one of the PaaS providers, based on the offers that were negotiated automatically by the Cloud4SOA system on his behalf. Tom would now like to deploy his application to the PaaS platform he has selected in the previous step. Therefore, he will use the appropriate UI in the Cloud4SOA portal to upload his application and give the command to deploy it to the PaaS platform. (2) **Check-State**: Tom Pérez deployed his solution to a PaaS environment. Tom would now like to check on the status of his application on the PaaS platform. So he will log-in to the Cloud4SOA portal and see if his application is still at "started" state. (3) **Un-deploy**: The solution Tom has developed has come to the end of its lifecycle. Tom would now like un-deploy his application from the PaaS platform. So he will log-in to the Cloud4SOA portal and give the un-deploy command.

No	Use Case Name	Short Description
3	**Monitoring**	Tom would like to check on his application on the PaaS platform. So he will log-in to the Cloud4SOA portal to obtain some monitoring data for his application.
4	**Accounting and Billing**	Tom would like to get some accounting and billing information for his application on the PaaS platform. So he will log-in to the Cloud4SOA portal, and activate the Cloud4SOA Accounting and Billing widget that provides him Accounting and Billing information about the service.
5	**SLA Violation**	The Cloud4SOA system monitors Tom's application, and at some point in time, the SLA enforcement module detects or preventatively foresees an SLA-Violation (i.e. it detects that the PaaS platform is not complying with the SLA terms, or it is close to not being able to comply with the SLA terms). Depending on Tom's SLA enforcement policy and configuration of the Cloud4SOA SLA Module, Tom's application may be given more resources to complete its task, moved to another resource or offered a discount in the bill. If the SLA breach is serious and cannot be recovered, the service may be terminated.
6	**Migration**	Tom is not very happy with the services and capabilities he is getting from the PaaS provider, so he would like to change the provider. He will log-in to the Cloud4SOA portal to search for a new PaaS provider, and migrate his application to the new platform he finds.

7. Governance Layer Architecture

The Cloud4SOA Governance Layer consists of four modules and seven components (the white boxes in fig. 11.5 below). Enclosed in parentheses is the WS Agreement specification nomenclature:

- **SLA Module**
  - SLA Negotiator Manager (agreement initiator)
  - SLA Negotiator (agreement factory) service
  - SLA Enforcement (agreement) service
  - SLA Decisor
- **Execution Management Service (EMS) Module**
  - Execution Management Service
- **Accounting and Billing (A&B) Module**
  - Accounting and Billing interface
- **Monitoring Module**
  - Monitoring interface

Fig. 11.5. Overview of the Governance layer components

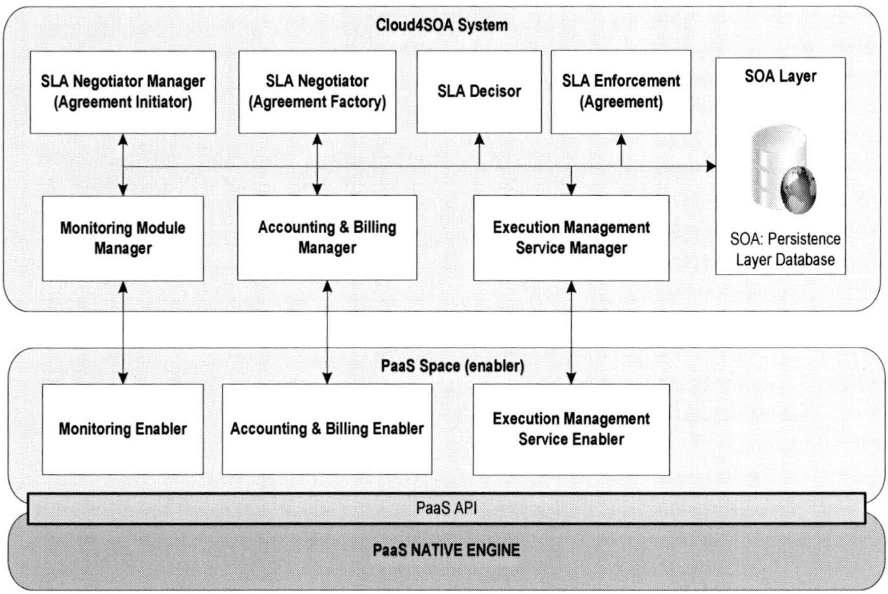

7.1 SLA Management Module

The SLA package is responsible for the SLA management activities which allow the establishment of an agreement between a cloud-based application developer and the corresponding cloud PaaS provider, the contracts enforcement taking into account monitoring information and the triggering of the appropriate recovery action when a violation is raised. SLA package is further broken down into three components:

- The SLA Negotiation
- SLA Enforcement
- SLA Decisor

7.1.1 Motivation for using a restful implementation of WS-Agreement

The Cloud4SOA consortium decided to rely on WS-Agreement specification as a standard for creating and monitoring Service Level Agreement contracts in distributed systems and WS-Agreement Negotiation for dynamic negotiation of Service Level Agreement contracts. We considered a few different WS-Agreement implementations as the baseline technology for the Cloud4SOA SLA Management framework. WS-Agreement for Java (WSAG4J) (https://packcs-e0.scai.fraunhofer.de/wsag4j/) is a generic implementation of the WS-Agreement protocol by SCAI. It supports common functionality to create and monitor agreements and enables users to build and deploy WS-Agreement based services. This implementation is based on Web Services Resource Framework (WSRF), which is quite heavy weight to manage, and it uses MUSE which is not so stable yet.

7.1.2 SLA Negotiation

The SLA Negotiation component in the Cloud4SOA Governance Framework provides capabilities to perform automatic negotiations on behalf of PaaS providers, based on SLA terms specified by the user. This component is broken down into two sub-components:

- SLA Negotiator Manager
- SLA Negotiator

Every PaaS provider might have a different format for SLA agreements, so first a template is retrieved by the SLA Negotiator Manager sub component, which represents the user, from the SLA Negotiator sub component, which represents the Provider. Then the template is filled to create an offer and presented to the SLA Negotiator. The SLA Negotiator either confirms this offer or returns a counter offer.

The Cloud4SOA SLA Management Framework is mainly based on the WS-Agreement Standard. In order to provide WS-Agreement capabilities on behalf of both the Cloud4SOA platform users and the PaaS Providers, the SLA Negotiator component is made up of two sub-components, one for the Agreement Initiator side (SLA Negotiator Manager), and one for the Agreement Responder side (SLA negotiator).

7.1.3 SLA Enforcement

SLA Enforcement is in charge of supervising that all the agreements reached in a contract are respected. It receives and processes all measurements related to the execution of a business application sent by the platform monitoring interface, and checks whether the measurements are within the thresholds established in SLA contract for QoS metrics. Whenever the execution of the business application does not satisfy these conditions, the SLA Enforcement notifies the SLA-Decisor service, which is in charge of deciding the appropriate recovery action to take.

The SLA Enforcement Component is started by the Execution Management Service, upon deployment of user application on the Provider's Platform. After deployment, SLA Enforcement regularly gets monitoring data from the platform and compares it with the SLA terms that were specified in the SLA Agreement between the user and the Platform Provider. An SLA Violation is raised to the SLA Decisor component, if the SLA terms are not met, as seen in fig. 11.6 below.

Fig. 11.6. SLA Enforcement and SLA Decisor Use case

7.1.4 SLA Decisor

When the application is deployed, the monitoring component periodically sends measurements to an SLA-Enforcement instance that, taking into account the SLA contract signed by the application consumer and PaaS provider, decides if the application is running in the expected conditions (QoS parameters within thresholds). If a violation occurs, this is managed by the SLA-Decisor which makes use of policies to take appropriate recovery actions. It receives and manages notifications from the SLA Enforcement only when an SLA contract is not respected and an adaptation strategy should be taken.

It can also act as Policy Manager to retrieve the suitable policy associated to the violation according to the execution context. This policy contains the actions that must be undertaken (i.e. show informational messages, increase processes priorities, apply discounts, destroy application pool, migrate application to another PaaS, etc), depending on the seriousness of the violation, the context of application and the overall status of the PaaS. In the first case, it may be only necessary to notify the EMS, which will apply corrective actions, whereas in the second case it recovers from the situation by applying global corrective actions.

The SLA Decisor service mainly interacts with the SLA-Enforcement Service and the Execution Management Service (See fig. 11.6 above).

7.2 Execution Management Service

The Execution Management Service (EMS) is the main interface to the PaaS platform in all things application lifecycle related. This mainly includes deployment, undeployment, maintenance and migration related tasks. It is separated into two main parts inside the Cloud4SOA system itself and in the Enabler. This design decision was made in order to support a wide variety of PaaS platforms independent of the PaaS provider's support. Internal research has shown that there is very little interest from PaaS providers at this time to support the Cloud4SOA system. While they see the benefit of easy migration potentially bringing more customers, they also see the risk of losing them. However, both effects are considered much too small by the providers as a financial incentive to justify the software development efforts needed. The chosen design of internal components which provide the essential functionality with an internal API and an Enabler implementing the internal API and the PaaS platform specific API solves this problem by making it easy for external developers to write Enablers for more and more platforms.

Figure 11.7 explains the basic flow of commands through the different stages of the system, starting at the SOA layer ending at the PaaS provider. Besides enabling the support of more PaaS platforms the separation of the EMS into two parts allows to support additional logic in the Cloud4SOA based EMS part. This could be used to support auto selection of Enablers depending on the needed functionality. One problem of the Enabler living in the actual application is that for first time deployments there is no application running the Enabler. This problem can be solved if Cloud4SOA knows other enablers for that platform exist and can use one of those to deploy this application. Logic like this can live in the Cloud4SOA part of the EMS while the actual translation of internal API calls to platform provider specific API calls is then handled in the Enabler. This introduces security concerns if the Enabler on the platform is modified to log credentials. Therefore, it is important that this process does not involve actual customer credentials going through the enabler.

Fig. 11.7. Execution Management Service interaction diagram

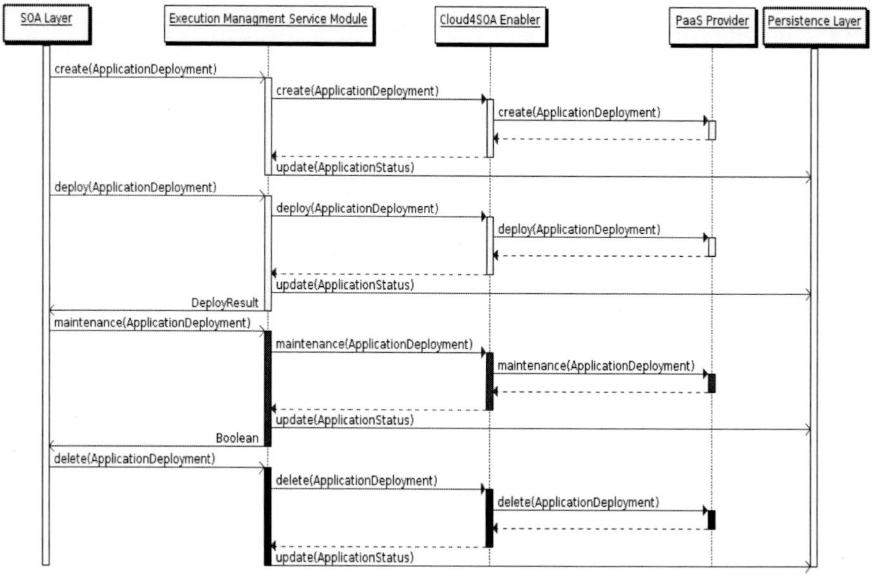

7.3 Platform Monitoring Interface

The Platform Monitoring Interface is in charge of periodically measuring application and platform health and storing the results for future analysis and use by other Cloud4SOA components, most importantly the SLA enforcement module. The architectural design is driven by the need for a very robust and scalable environment as outlined in the sequence diagram. A major challenge for the monitoring functionality is the unavailability of certain measuring data needed by the SLA Enforcement Module as specified above. It is highly dependent on the ability of the Enabler to get the data from the application's underlying PaaS platform either by natively requesting that data from a provider specific API or by collecting that data through Enabler internal code. If certain information can't be obtained by either of the two methods from the Enabler this could result in reduced availability of monitoring and as a result SLA enforcement capability of Cloud4SOA for that specific PaaS platform.

The Platform Monitoring Interface is divided into different components, mainly the Job Interface and the Job Workers which asynchronously poll application and platform health information and store them in the Cloud4SOA Persistence Layer. This asynchronous design allows the

Platform Monitoring Interface to scale seamlessly to large numbers of monitored applications and platforms because the workers can be easily distributed in a horizontal fashion. Workers poll application and platform health information by sending requests to the platform specific Enablers. The details of which information can be collected by the Enabler and requested by the monitoring workers heavily depends on the design of the Enabler and the platform specific functionalities and available APIs.

Fig. 11.8. Monitoring interaction diagram

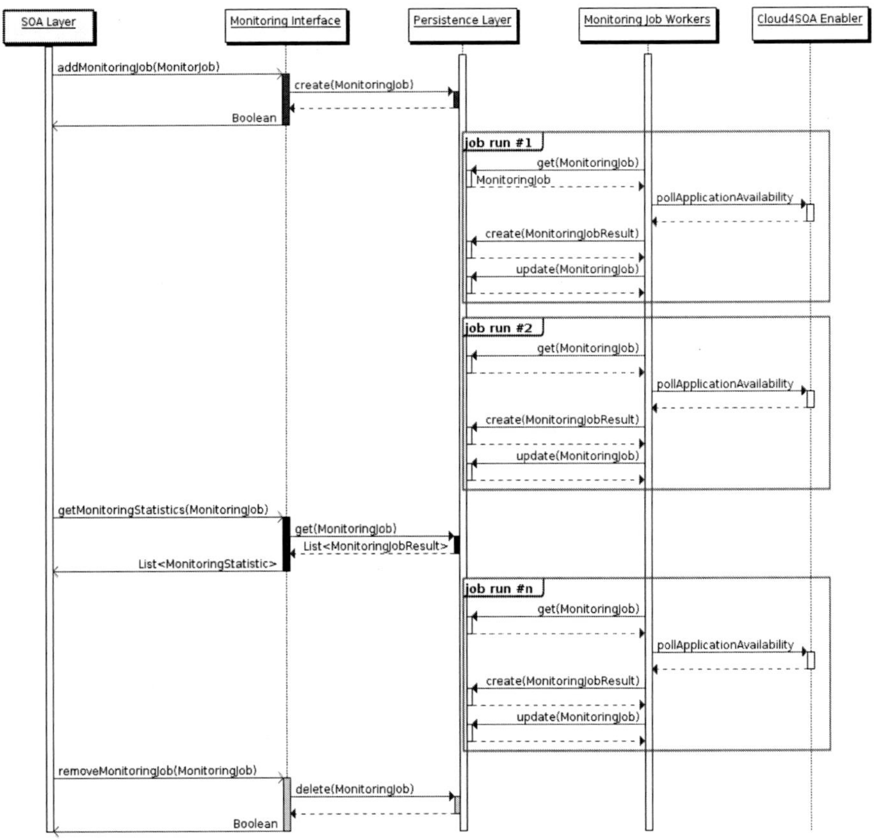

7.4 Accounting and Billing Interface

The Accounting and Billing Interface is a very basic but important functionality, since price is always a critical factor for business decisions. While the functionality is not problematic per se, what is problematic is the huge differences in how different providers charge customers. The goal of the Accounting and Billing Interface design is to show the most information possible, but falling back to the minimum measurement of money per time. One of the most important uses for the Accounting and Billing Interface and its data is the use case of comparing two different providers. Here the collected data can be used in a very innovative way. Anonymous billing data from similar applications on other providers can potentially be used to make recommendations to Cloud4SOA users based on their application's profile. This, however, is not part of the Accounting and Billing Interface itself—it merely enables this functionality through its data collection features.

Fig. 11.9. Accounting and Billing interaction diagram

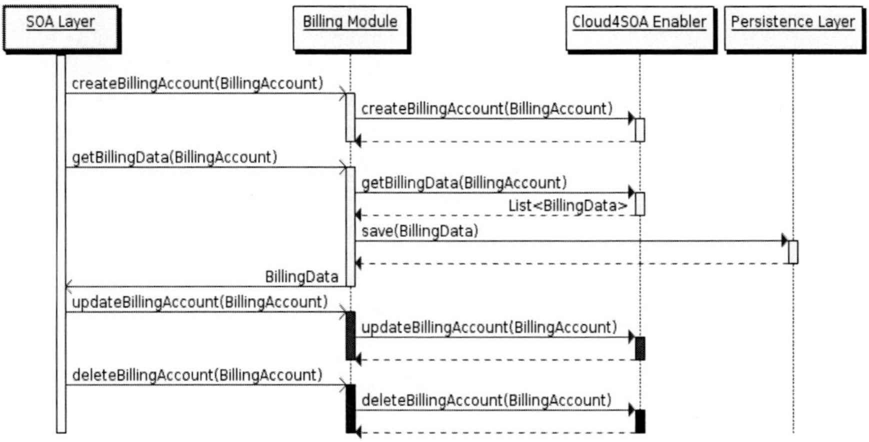

Most of the Billing Module's methods are a simple CRUD interface. The translation between Cloud4SOA internal API and platform specific API is done in the Enabler.

8. Business Benefits

The Service Lifecycle Governance Framework is an integral part of the Cloud4SOA integrated system's ability to facilitate interoperability and portability functionalities between PaaS offerings of shared technology. Application portability, in particular, is a chief adoption barrier among potential PaaS users, echoing the similar vendor lock-in interoperability concerns of the IaaS market segment. In the backdrop of the vendor lock-in risk is a landscape of highly fragmented and specialised PaaS offerings.

The innovation potential of PaaS is not just becoming recognised by the large established cloud and middleware providers, but also represents an entry point for providers of smaller size, such as Europe's SMEs, where the large resources needed to operate data centers is not a prerequisite for hosting your platform on top of external infrastructure.

In this context, the capabilities of Cloud4SOA's reference architecture, framework and resulting integrated system can help open up this market to smaller competitors; not only aiding in announcing their characteristics of searching for adopters, but also facilitating the jump from one platform to another for existing application developers. Although at first this could be seen as a vulnerability to a PaaS provider's developer base, the benefits of initiatives like Cloud4SOA can help grow the pool of adopters, as well as create added value and a competitive advantage for those platforms participating. Governance issues, such as platform selection, negotiation, provisioning, operation, and application runtime management, are again key characteristics to develop and implement in a system aimed at these challenges.

To help test and evaluate the project's technical outcomes, Cloud4SOA will also run three showcases: business intelligence with Portugal Telecom, network monitoring with RomTelecom, and business cooperation with Fraunhofer FIT's BSCW platform. The project will assess the performance of these showcase prototypes through the analysis of result testing, and generate the lessons learned (from their development, deployment and execution), formulating them as a step-by-step cookbook and methodological adoption guidelines.

External feedback and awareness of the cloud industry's reality is also critical when dealing in the young and constantly evolving PaaS market segment. To support the project in this environment, Cloud4SOA hosts a Special Interest Group, made up of PaaS and IaaS experts, cloud application developers, vendors, standards development representatives and academic researchers.

9. Conclusions

This chapter has presented a Reference Architecture, an outcome of the Cloud4SOA research project, which aims to resolve semantic interoperability issues that arise between heterogeneous cloud computing platforms sharing the same technology. The chapter introduced a new approach for administrating business applications and managing SLAs in a PaaS context through the strict relation between the SLA Management and the PaaS Governance. Our next step will be to complete the first simple implementation of the proposed architecture, and validate it in real business situations.

Acknowledgement

This work was partially supported by the European Commission through the Cloud4SOA project (FP7 - 257953).

References

D'Andria, F. "An Enhanced Strategy for SLA Management in the Business Context of New Mobile Dynamic VO." In *Proceedings of the eChallenges 2006 Conference (eChallenges 2006)*. Amsterdam: IOS Press, 2006.

D'Andria F. "Dynamic SLA and Trust for Next Generation Business Models in Grid". In *Proceedings of the eChallenges 2008 Conference (eChallenges 2008)*. Amsterdam: IOS Press, 2008.

CoreGrid. A Comparison of SLA Use in Six of the European Commissions FP6 Projects, 2007. http://www.coregrid.net/mambo/images/stories/TechnicalReports/tr-0129.pdf (accessed August 11, 2011).

Kübert, Roland, Katsaros, Gregory and Wang, Tinghe. "A RESTful implementation of the WS-Agreement specification." In Proceedings of the Second International Workshop on RESTful Design, WS-REST, 2011.

Krummenacher, R., Hepp, M., Polleres, A., Bussler, C. and Fensel, D. "Www or what is wrong with web services." In: *ECOWS '05: Proceedings of the Third European Conference on Web Services*. Washington DC: IEEE Computer Society, 2005.

Fensel, D., Lausen, H., Polleres, A., de Bruijn, J., Stollberg, M., Roman, D. and Domingue, J. *Enabling Semantic Web Services: The Web Service Modeling Ontology*. New York: Springer, 2006.

Jacek K. et al. "SAWSDL: Semantic Annotations for WSDL and XML Schema." *IEEE Internet Computing* 11 (6) (2007): 60–67.

CHAPTER TWELVE

REDUCING TIME TO MARKET WITH THE PLATFORM AS A SERVICE CLOUD OF THE FUTURE

JOSE LUIS VAZQUEZ-POLETTI, EDUARDO OLIVEROS, RAFAEL MORENO-VOZMEDIANO, IGNACIO M. LLORENTE, SEBASTIAN ORTEGA, MIGUEL JIMENEZ, JAVIER SORIANO AND ANDREAS MENYCHTAS

Cloud computing is essentially changing the way services are built, provided and consumed. Despite simple access to clouds, building elastic services is still an elitist domain and proprietary technologies are an entry barrier especially to SMEs. It therefore remains largely within the domain of established players. This chapter will explain how the 4CaaSt project will bring significant benefits to the developers of cloud-scale applications to the European economy, by providing an easy to use Platform for a more competitive environment, greatly simplifying design and delivery of tailored services and compositions.

1. Introduction

Standardisation of IT infrastructures, especially through virtualisation, has led to significant deployment and management savings in time, money and accelerated innovation. On the infrastructure level, resources are available virtually infinitely and consumed according to the current demands instead of allocation with a long term horizon. However, this elasticity has not made it to a broad set of applications and application providers who are left alone when providing elastic applications. Providing and operating scalable services still requires complex IT landscapes and is an expert domain. It requires high budgets with sometimes large upfront investments and highly skilled IT staff, thwarting especially small and medium enterprises providing applications (OSGi, JSR220, JSR 154) (Banga et al. 1999; Herzog and Shahmehri 2005; Czajkowski and Von Eicken 1998).

1.1. Motivation

The motivation of 4CaaSt is to build a platform which lifts the properties of elasticity, availability, and pay-per-use characteristics up to the platform level, forming the foundation for the efficient execution of elastic applications by medium skilled application providers.

More specifically, the construction of scalable distributed applications is still an unattainable issue for most market players (especially SMEs). This task involves a very detailed knowledge of how to deploy an application in tiers and make each of those tiers scale as needed by the services themselves, not only by the hosting infrastructure. Moreover, large corporations have skilled personnel specifically for these tasks, for example mapping a coded application into an actual production environment. These skilled personnel are hard to find and often unaffordable for SMEs or large companies without the required technical background. At the same time, developing such applications without a staff of expert programmers in distributed systems is often close to impossible for small corporations due to technical and budgetary reasons. This traditional separation of roles results in high operational costs. While keeping the roles separate is still critical (developers should focus on developing, deployers on deploying, and infrastructure operators on maintaining data centers and keeping networks up and running), 4CaaSt envisions that some of the responsibilities of these roles can be automated and therefore offered "as a service" to help other roles to focus on their core business (Merrill Lynch 2008; O'Reilly 2009). This already difficult picture is further complicated if the application provider aims at having a "smart" application that scales vertically or horizontally according to actual demand.

4CaaSt makes a distinction between elasticity and scalability. While the latter is focused on run-time adjustment of resources and implies more infrastructure-oriented metrics, rules to scale, and resultant actions that are far below the mindset of application providers, the former faces scalability in a more business-oriented context. This implies that the metrics and rules to scale are closer to application providers, letting them forget about infrastructure terms, and focus on their actual business: developing applications people may be interested in.

PaaS Clouds aim to bring cloud benefits closer to SMEs and non technical people, so they can deliver their applications on an automatically scalable platform that offers ease of service definition, development, marketing, delivery and maintenance. This way, value added services can blossom in a highly competitive environment.

4CaaSt envisions this elasticity at two different levels:

- **Computing level**, which is beyond 4CaaSt, but controlled by 4CaaSt by means of IaaS-enabled scalability rules, for example combining Amazon Auto Scaling (https://aws.amazon.com/autoscaling/) and Amazon CloudWatch (https://aws.amazon.com/cloudwatch/), RightScale scaling rules (http://www.rightscale.com/), or RESERVOIR's elasticity rules (Amazon B. Vaquero et al. 2009; Rochwerger et al. 2009; Leymann 2009).
- **PaaS elasticity**, the PaaS platform will be in charge of reshaping the application tiers as needed in accordance with some user-provided application level metrics.

The computing level (IaaS) must be abstracted in order to outsource the roles and functions enabling cloud deployment and operation. These responsibilities should be assumed and encapsulated by the PaaS, whose API is made up of a series of execution environments, added-value services and infrastructure utility services.

Ease of use, flexibility and abstraction gradient of exposed services are critical adoption factors for SMEs and non-technical users, otherwise outside of this market.

1.2. Clouds as ICT infrastructure utilities

Current clouds represent a major step in the evolution of hosting. They enable the provision of ICT Infrastructures and/or basic Platforms "as a Service," exhibiting many of the features provided by classic utilities, namely:

- A very simple interface that enables self-provisioning and monitoring of resources, hiding underlying complexity
- Pervasiveness (the service is accessible from almost anywhere)
- Allocation of resources is activated and can scale "just in time"
- Price per unit of consumed resources is rather competitive
- Users pay per actual and effective usage of allocated resources

- The service is reliable and secure.

Relying on the utility-like features that clouds provide for hosting, application providers can minimise their CAPEX and OPEX on ICT infrastructure required to run their applications, without dealing with all the burdens of managing and operating that infrastructure (Thanos et al 2009).

The definition of cloud computing provided by the National Institute of Standards and Technology (NIST) differentiates the following three cloud service models with respect to the capabilities provided to the service consumer and the location where the cloud infrastructure is operated (Mell and Grance 2009); fig. 12.1 illustrates these three cloud deployment models:

- **Cloud Software as a Service (SaaS)**: The capability provided to the consumer is to utilise the provider's applications running on a cloud infrastructure. The applications are accessible through a thin client interface such as a web browser. With the possible exception of limited user-specific application configuration settings, the consumer does not manage or control the underlying cloud infrastructure or even individual application capabilities.
- **Cloud Platform as a Service (PaaS)**: The capability provided to the consumer is to deploy consumer created or acquired applications onto the cloud infrastructure. The consumer does not manage or control the underlying cloud infrastructure, but has control over the deployed applications and possibly application hosting environment configurations.
- **Cloud Infrastructure as a Service (IaaS)**: The capability provided to the consumer is to provision fundamental computing resources such as processing, storage, or networks where the consumer is able to deploy and run arbitrary software. The software can include operating systems and applications. The consumer does not manage or control the underlying cloud infrastructure but has control over operating systems, storage, deployed applications, and possibly limited control of select networking components.

Fig. 12.1. Cloud applications and cloud types

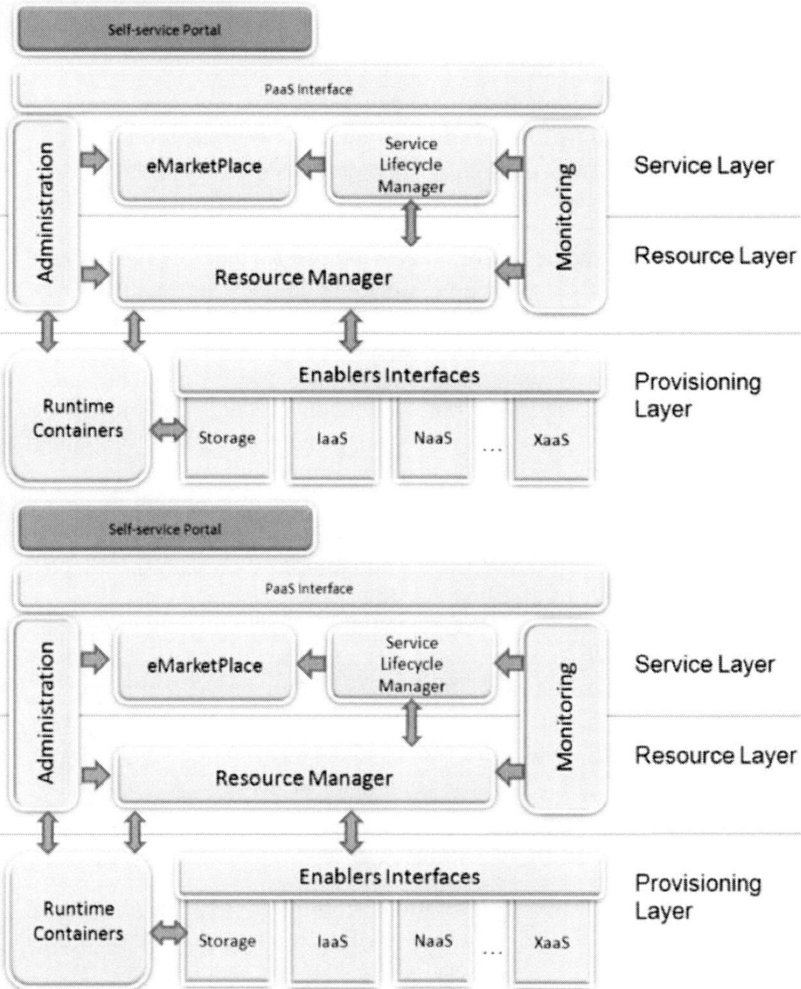

Although current IaaS Clouds have proven mature enough to be released to the market, a paradigmatic example such as Amazon still leaves application providers almost alone in managing the software stack of their virtual machines and addressing the elasticity of their applications. In those platforms, it is the application provider's duty to develop and arrange the developed software in an appropriate manner for a real production environment (to set up different tiers, configure load balancers, data distribution, replication, authentication, billing, etc.). While load-balancing and auto scaling is available, it comes at the price of steep costs of additional instances. In other words, current IaaS Clouds are too infrastructure oriented.

Current representatives of PaaS Clouds, such as Google App Engine, go beyond this. However, it suffers from limitations such as the inability to be used for private PaaS within a company, a lack of defining storage locations for human resource data, for example, and a "lock-in" effect due to the missing bulk up-/download features. This is, indeed, a serious limitation for the SaaS model to overcome. In this regard, we envision that the natural evolution of current PaaS Clouds comprises the inclusion of new capabilities in the "platform" on top of which applications run, related to dynamic application deployment, such as programming models for Internet-scale applications with integrated higher level Telco services including application interfaces for innovative convergence scenarios. We further envision an advanced lifecycle management that simplifies and reduces the cost of service variants and compositions (Gilles et al. 2009).

Last but not least, a convergent allocation of computing and communication resources is not properly solved in current IaaS/PaaS clouds. Elastic allocation of the necessary computing resources will not be enough to comply with established Service Level Agreements (SLAs). Thus, for example, an application may claim a higher bandwidth assigned to outbound connections in order to support some peaks in demand, or it can ask for the creation of Session Initiation Protocol (SIP) sessions among users, etc. Therefore, it is envisioned that future PaaS should be implemented on top of both IaaS and advanced NaaS (Network as a Service) capabilities.

1.3. Clouds as business ecosystems

Clouds host execution of many applications from multiple providers. Our vision is that clouds will naturally evolve to become worldwide catalogues/repositories of cloud services and cloud applications. This circumstance places cloud providers in a privileged position to act as intermediaries between Service Providers which host applications on their clouds, and Service Consumers of those applications.

In this respect, we may anticipate that the natural evolution of current PaaS Clouds comprises the inclusion of new capacities in the "platform" on top of which applications run, related to marketplace features. In that sense, the PaaS layer will include techniques for effective development and adaptation of the applications with respect to their business aspects.

The Service Provider would take care of billing for the usage of applications registered in the cloud. This would typically imply that the platform implements features for multiple billing and payment mechanisms as well as real time rating/charging and application usage accounting (Elmroth et al. 2009a and 2009b), which are rather complex features to be implemented by many Service Providers, especially small-to-medium Service Providers, on their own. Service Providers would manage revenues and perform the proper settlement with application providers in a regular basis (monthly, for example). It is worth mentioning that the described "clearing house" function may be attractive enough for cloud providers to be offered even to Application Providers who may have decided to host their applications on other clouds or on their own premises. Federation is then a key feature, enabling a given cloud provider to intermediate access to a large number of applications from different application providers, some of which will run over the cloud infrastructure/platform, while others will not (Stanoevska and Wozniak 2009; Rochwerger et al. 2009). 4CaaSt aims at creating such an scenario of federated cloud instances, both public and private, that can benefit from each other. Following the SPI (Service, Platform, and Infrastructure) cloud model (Mell and Grance 2009), cloud computing products are categorised in the three main layers of Infrastructure as a Service (IaaS), Platform as a Service (PaaS) and Software as a Service (SaaS). Lately, more and more "buzzwords" for cloud products have appeared describing IT assets that are offered as services such as Database as a Service (DaaS), Context as a Service (CaaS) and other; all wrapped under the term Everything as a Service (XaaS, EaaS or *aaS) (Lenk et al. 2009). What all these products have in common is the need for marketplaces and platforms capable of supporting the technical and business requirements of providers and consumers, allowing them to exploit the technological advantages of clouds to expand their businesses and minimise their costs.

The existing cloud solutions come with several limitations regarding the involvement of application stakeholders in the cloud ecosystem because of their various, often complex and contradicting, business and technical requirements. In addition, there is considerable overhead for third parties to deploy their applications in the cloud infrastructures, create new business models and establish synergies, since there are many dependencies between the services which cannot be fulfilled by a single provider. In the mobile phones paradigm there are several approaches addressing this problem, the most well-known and successful being the iPhone App Store and Android Market. Developers and providers participate in these marketplaces selling their applications and services using various business and revenue models. These solutions have led many developers and providers to be involved, extending their businesses in the mobile market, while end users are now able to discover hundreds of services and applications that satisfy their needs.

The adoption of clouds as technologies and solutions for developing, providing and consuming services and applications for IT is continuously increasing. Similarly, more and more organisations, enterprises and SMEs as well as individuals are involved in this computing paradigm, exploiting its benefits and extending its functionality at all layers. As already mentioned, the concept of cloud computing extents from the hardware resources layer to the delivery of complete applications, thus several types of actors participate in the processes taking place in each layer. In that sense, the service delivery process becomes more complicated and spans different administrative domains, leading to multi-part value chains and more complex lifecycles for applications and services.

The simplified model of provider-consumer trading resources or services is not applicable in modern clouds where any kind of IT assets can be offered either independently or through compositions. At business level, cloud marketplaces offer the functionality of providing/consuming composite applications. The marketplaces play the role of broker between providers and consumers, and additionally allow service aggregators to participate in the value chain by developing new products. This concept is expected to minimise the business complexities for providing and consuming all types of services by supporting all phases of the service lifecycle (knowledge, intentions, contract and settlement). The marketplaces offer the providers the ability to publish services and applications in a managed environment, which controls the business terms and conditions (price, revenue sharing, promotion, etc.), including integrated rating and billing capabilities. Selection of products driven by customers' needs is another important aspect of the marketplaces in order to allow efficient balance between cost and QoS required at business level. As with all modern IT environments, clouds should be capable of supporting product customisations that cover the particular needs of each customer and at the same time provide mechanisms and tools to the developers to develop new products or reuse existing ones in compositions. In that sense, automated selection processes, not only of products but also of pricing models and SLAs through which these products are provisioned, will be part of the core cloud marketplaces' functionality. Therefore, the resolution of the customers' requirements becomes a key process for marketplaces, especially if we consider that cloud marketplaces are capable to offer any kind of product (XaaS), each of which can be offered through different pricing schemas and SLAs.

Another important aspect of the future clouds to be addressed by the marketplaces has to do with the concept of "Mashup as a Service" (MaaS), resulting from the integration of the Web 2.0 mashup and social web concepts (Gilles et al. 2009), enabling the creation of a community of developers and users around the cloud.

In summary, support of marketplace functions as well as MaaS capabilities will provide the answer to questions application providers would find hard to solve on their own: how can they monetise applications/services, and how can they better promote them (Thanos et al. 2009; Assuancao et al. 2009).

1.4. Clouds as rich programming frameworks

Early PaaS platforms sacrificed functionality to attain scalability and elasticity. They provided rudimentary storage APIs, simple libraries and containers to build applications, etc. Recent developments in the PaaS arena such as Windows Azure (http://www.microsoft.com/windowsazure/) and WebSphere CloudBurst (http://www-01.ibm.com/software/webservers/cloudburst/) are changing that landscape. Windows Azure provides a wide programming framework for Cloud computing including .NET framework (application server), SQL Azure (relational database) and a myriad of Web services to support interaction across application components. CloudBurst is the "Cloudification" of the IBM WebSphere service platform which also has a rich programming framework. Both are examples that announce the race that is about to start: a race between cloud providers towards providing the richest programming framework for application providers.

The winners of this race will dominate the future of the cloud scene. However, the lack of standards in this area will compromise the portability of applications across different clouds. The risk of applications being locked in to specific cloud providers is very high. The fact that current PaaS initiatives are dominated by US companies leads us to the conclusion that Europe must not lose the initiative if it wants to play a strategic role in the future of cloud computing. 4CaaSt is aiming to have a pivotal role, addressing the open specification of a relevant number of base programming libraries in the cloud, covering functions such as:

- Access to Context Services, considering Context in its widest sense (location, delivery platform info, user profile, info about social network the user belongs to, etc.)
- Advanced Publish/Subscribe (P/S) services, allowing for smart, effective, context-aware and scalable content-based addressing communication channel.
- Access to Network Enablers (for sending SMS/MMS, establishing SIP sessions, location services, etc.)
- Access to advanced cloud databases and data stores.
- Advanced monitoring services enabling real-time and asynchronous management of cloud resources.

These services conform the aforementioned PaaS abstraction and, more importantly, embody the 4CaaSt vision. For example, the P/S service provides a communication pattern that is typically leveraged as part of complex distributed systems and plays a supportive role both within the platform and to the hosted components.

PaaS offered services should be as reusable and versatile as possible, and the 4CaaSt response to this issue relies on context-awareness handled by a context-broker offered as a service that potentially permeates the whole PaaS. Consequently, any service can consume context information in a homogeneous and decoupled manner so as to complement its service interface and become more flexible and reusable. Apart from the reusability advantages, context-awareness can foster performance by dealing with proximity, similarity, ubiquity, etc.

Many of the PaaS services depend on network communication capabilities. Our European Telco Service Providers are strong global players who will be in a strong position to lead standardisation. Application Providers in Europe, for example ISVs such as SAP or integrators like Bull, are also strong globally and can bring the necessary expertise in business critical applications to drive the definition of significant APIs. Through 4CaaSt, Application Providers and Telco Service Providers can join forces to position Europe in the "race for the clouds."

2. The 4CaaSt Vision

The 4CaaSt project aims to create an advanced PaaS Cloud platform which supports the optimised and elastic hosting of Internet-scale, multi-tier applications and embeds all the necessary features easing programming of rich applications and enabling the creation of a true business ecosystem where applications coming from different providers can be tailored to different users, mashed up and traded together.

The 4CaaSt platform therefore brings a unique value proposition, translated into the following points:

- The highest level of abstraction regarding hosting, hiding complexity and easing the management of applications in the cloud.
- A complete set of built-in programming libraries and common facilities beyond what is offered by state-of-the-art PaaS clouds, easing development of smart cloud-scale applications
- The most attractive business ecosystem support facilities, to promote and monetise applications as well as create an active community of users and developers

- The necessary tools to monitor execution and manage lifecycle of applications.

Overall, the project will bring significant benefits to the European economy via a greatly simplified design and delivery model for services and service compositions. The developed platform will lead to the establishment of new and highly dynamic and innovative service ecosystems.

The project will bring specific benefits to the following stakeholders:

- Application providers, who get a platform that allows for easy development and deployment of Internet-scale services as well as advanced business and pricing model for merchandising each application. They can focus on the service functionality and can use platform functionality and underlying resources without any a priori capital expenses.
- Consumers, who get easy access to a large number of services with very easy-to-use consumption models (pay per use, for example) based on their individual technical and business requirements.
- Application/Service aggregators, who get a platform that allows for easy composition of services which all follow the same lifecycle model.
- Cloud providers, who can establish a platform in order to establish a business/service ecosystem around a certain topic and to generate a business on intermediation.

To summarise, the platform should enable the involved parties such as Service Providers and Service Consumers to focus on their individual business.

3. Architecture

To realise the 4CaaSt vision, the project defines and implements a novel architecture that integrates the ideas and concepts of the 4CaaSt framework.

3.1 Architecture Components

As shown in fig. 12.2 below, 4CaaSt is organised in a layered structure. The 4CaaSt approach supports the use of a set of heterogeneous underlying technologies.

Fig. 12.2. 4CaaSt overall architecture

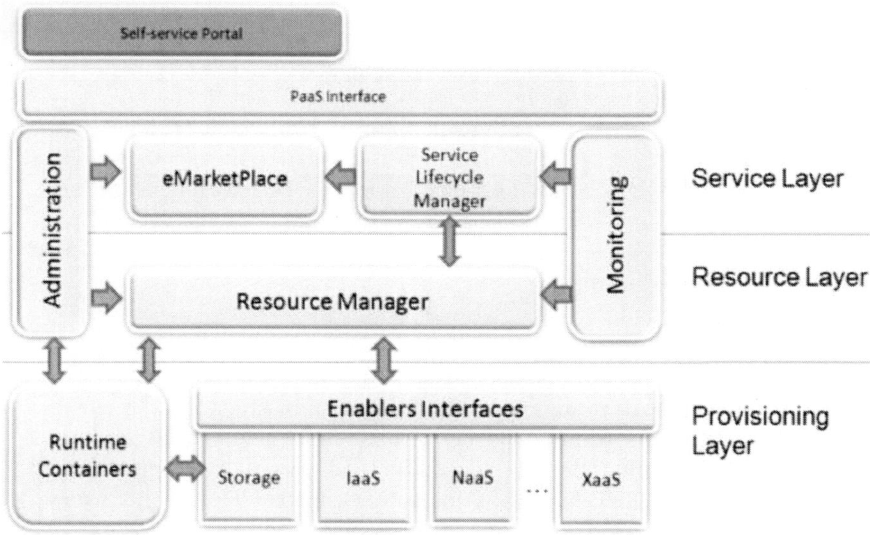

The topmost layer, the Service Layer, is in charge of interacting with the end-users of the 4CaaSt framework through a Web GUI and a public PaaS API. The services marketplace is in charge of managing the commercial offers in the 4CaaSt platform, from its definition to the settlement and billing processes. The Service Lifecycle Manager is in charge of governing the deployment process (from resolving the technological requirements to asking to the Resource Layer to provision the required execution environment) of the Services. The Service Layer also integrates the administration and monitoring functions offered to the end-users.

The Resource Layer is in charge of providing execution environments and managing the deployment of applications on them. This layer calculates the amount of virtual resources needed to host the execution environment, and controls the scalability of them. The execution environment is tailored to the virtual infrastructure with the aim of increasing the overall platform efficiency

The Provisioning Layer comprises capabilities for accessing virtualized resources provided by IaaS (Virtual Machines & Storage), NaaS (Virtual Networks & Networking elements) and virtualised services provided as a Service (XaaS). This layer hides the details of how virtualised resources are provisioned, enabling the integration of different providers by adding new Enabler Interfaces.

3.2 Enabler interfaces to native technologies

Enabler interfaces are implemented by pluggable modules in order to provide an infrastructure service to cloud-hosted components in a vendor-independent fashion. As a consequence, enabler interfaces must define standardised, generic services and must be tailored for the cloud environment. These premises turn enabler interfaces into unique selling points of 4CaaSt's architecture; since they offer homogeneous and simple APIs to fully exploit cloud technologies, providing great added value in several functional domains.

3.2.1 Context-awareness in native cloud APIs

Cloud-native context-awareness could bring optimisations and improve user experience at different levels from infrastructure provisioning, topology optimisation, interface adaptation, and even at the provisioning level (by choosing the services by the active context). However, current PaaS solutions lack a global context-awareness strategy and framework.

The 4CaaSt proposal for context-awareness is based on two premises: a global context broker offered as a service in order to provide a low entry-barrier, a homogeneous interface to produce/consume contextual information, and an architecture in which every layer and component (such as enabler interfaces, service lifecycle manager, and so on) is context-aware.

Thanks to built-in context-awareness on 4CaaSt architecture, it is truly possible for the PaaS user to create and deploy deeply context-aware applications. Otherwise, a context-agnostic PaaS makes it impossible to exploit context on tasks that are responsibility of the platform no matter what the application tries to do.

3.2.2 Content-based Publish/Subscribe enabling the event-based programming paradigm

A simple, distributed, and highly scalable communication mechanism among entities hosted in the cloud is needed. Publish/Subscribe (P/S) has been defined in the literature as a communication pattern that decouples source and origin in terms of time, space, and synchronisation (Eugster et al. 2003) and is regarded as an ideal mechanism for the exchange of contextual and other information.

Traditional P/S systems are topic-based, a simple yet powerful technique based on subscribing to specific labeled “channels” that has some limitations. When interests are diverse, either the number of topics required to keep the signal-to-noise ratio is overwhelming or the network overhead is impractical.

The proposed cloud-native technology, as part of 4CaaSt, leverages Content Based Publish Subscribe (CBPS) (Carzaniga et al. 2001; Jacobsen 2010; Jerzak and Fetzer 2008), an answer to overcome these shortcomings. Instead of subscribing to a set of topics, subscriptions are expressed as a pattern to be matched against published messages, so the topic-based distribution is replaced by content-based routing. CBPS releases the publishers from the duty of predicting the topics that could be relevant to subscribers. Moreover, message filtering can be distributed across a delivery network minimising network usage and enabling scalable broadcast-like distribution.

As previously stated, P/S must exploit context but current CBPS systems (Cugola and Munoz 2005) does not provide flexibility and context-awareness at the same time. Our proposal builds on the established CBPS APIs, adding new primitives to explicitly manage content in which the global context as a service interface can plug as a mediator. This enforces the separation of concerns since the context service can deal with context reasoning, mediation and homogenisation while the P/S service specialises on the context-aware routing of notifications.

3.2.3 MaaS-enabled business ecosystems

The service ecosystem proposed by 4CaaSt can be exploited through different front-end styles. On the one hand, an application provider can create value through a specific, complex and rich front-end to be used as a closed product exploiting their own or third party services. On the other hand, a service provider or an advanced end-user can offer a remixable front-end to exploit the advantages of end-user front-end web mashups. Both models are compatible and capture complementary market needs.

In order to lower the entry barrier to the consumption of services through front-end mashups, both the components and their integrated interfaces should be offered as a service of the PaaS, that is Mashup as a Service (MaaS).

MaaS relies on the usage of visual tools and portals supporting the creation of mashups. Using visual tools, end users may create backend mashups which orchestrate a number of existing services. Through Web portals, in turn, the end user is also capable of accessing a catalogue of available resources (gadgets) linked to different applications and relying on different backend services. In this catalogue, off-the-shelf components can be located and selected to be incorporated in a personal workspace, following a self-service paradigm. Then, the user can compose the selected gadgets to configure a final application to their particular needs without requiring any programming knowledge. Last but not least, the users can share information and knowledge about backend mashups they build, gadgets they use or the workspaces they setup.

3.2.4 4CaaSt marketplace

One of the fundamental objectives of 4CaaSt Project is to transform the clouds to open successful business ecosystems, an objective that is addressed by the 4CaaSt marketplace. The 4CaaSt marketplace will ease and enhance the communication between the various stakeholders involved in the service delivery process allowing for product compositions and complex business and pricing models. Until now, the majority of the cloud platforms have focused on the technical and technological aspects of clouds. In 4CaaSt, we found that the cornerstone for the wide adoption of clouds is native support of the business terms and conditions for providers, consumers, and aggregators. The marketplace will deliver tools and mechanisms to providers and aggregators to create any type of products based on the XaaS notion, which will be deployed and provisioned by the 4CaaSt platform or any other 4CaaSt-complaint cloud. In addition, these products will be offered under differed business and pricing models in order to cover all market trends or even create new ones. To this end, business simulation tools will be available to providers and consumers in order to develop effective models for the products or identify the business and operational parameters that will satisfy their particular needs, and will also maximise their profit.

The overall service lifecycle will be supported and supervised in business level from the marketplace. The main marketplace processes are categorised in five phases (fig. 12.3 below).

Fig. 12.3. Main phases of the 4CaaSt marketplace processes

Information: The information phase includes all processes and functionalities for providers to describe the business terms and conditions for the offered products and for customers to find suitable products and customise them. The processes will take into consideration the user profiles of the stakeholders in order to enhance the process, as well as the social aspects and feedback provided by other users for the particular product.

Business Resolution: One of the most advanced functionalities of the marketplace is business resolution. In the business resolution phase, the marketplace analyses the specific requirements of each customer, the technical and business constrains from the involved parties and identifies the best products combination under the certain circumstances including pricing models and SLAs.

Contracting: In the contracting phase the selected combination of products, pricing models and SLAs will be contracted, defining the obligations from all parties and the cash flow for the delivery of a product.

Settlement: During the settlement phase, the technical part of the contracted product as well as the technical terms of the SLAs will be relayed to the 4CaaSt Platform deployment services for initialisation, deployment and provisioning. Part of this phase is the continuous monitoring and management of the product following predefined policies, generic (scalin, for example) or product specific, that ensure the provisioning of the product on the required and contracted QoS level.

Analytics: In analytics phase, the performance of the product and product resolution is monitored and evaluated so as to support on the one hand the service providers to better define products, and on the other the customers to make better customisations and selections based on their particular needs. The analytics functionality of the marketplace is exploited in all other four faces allowing for more effective and optimised operation.

4. Current Status

The project is currently working on the first prototype implementation of the 4CaaSt platform. This prototype will focus on an initial set of features of the platform, like marketplace publication and contracting of applications, initial blueprint design, application deployment in the platform, initial integration with the cloud services and the monitoring system. More advance features like automatic elasticity, multi-tenancy support, complex network configuration or advance processing of monitoring data for analysis and recommendations will be the focus for the next years.

Two applications are used to show the 4CaaSt innovations after the first year. The first uses context information for booking the closest taxi, while the second application provides tourist information to the user while moving in the taxi. This application will integrate different cloud services like publish/subscribe, storage service for storing the information shown to the user, context-awareness to provide them with the most relevant information, and compound through a configurable mashup-based interface to 4CaaSt hosted applications and external services.

Three new scenarios will be developed during the next years to show the benefits of 4CaaSt to three market segments: SMEs, large corporations and the mass market. Each scenario will covert different features and innovations of the 4CaaSt platform, focused on the specific benefits that the platform provides to those market segments.

The Application Blueprint is a central element in 4CaaSt that contains the complete description of the application, including: application topology, requirements toward the platform and other technical information needed for correct deployment, and business aspects like price or pricing models to be used, SLA terms, etc.

A first specification of the Application Blueprint is available, but this specification will probably evolve during the project duration as new requirements appear from other parts of the platform, such as the marketplace, the PaaS management, scalability, monitoring, etc.

The project will work on the set of tools that will allow the creation of the Application Blueprint by developers and engineers for describing their applications, and the APIs required to interact with other parts of the platform but more significantly with the marketplace and the resource deployment and management layer.

The marketplace interacts with three types of users: the operator of the marketplace, the service provider, and the customer. The operator is allowed to create and manage the marketplace's users. The service provider is in charge of publishing new products in the marketplace, and the customer will select and contract a product.

When the service provider creates a product in the marketplace he or she adds the configuration that defines this product to the technical description of the application

A customer will then search over the product offer in the marketplace, and after selecting a product will define certain parameters of the application (usually related to the quality of the service). This process will instigate the resolution phase that will provide a final price to the customer. Finally, the customer can contract the application, and this will eventually launch the deployment of the application on the platform.

The marketplace will be extended in the future to support new functionalities like adding social network aspects (i.e. product evaluation or customer's friends suggestions), the billing and revenue sharing among the participants in the application execution, and simulation and business intelligence (for evaluating different pricing models and revenue sharing scenarios).

Planning the deployment of the application is based on the Abstract Resolved Blueprint (ARB) provided by the Service Lifecycle Management. The Deployment Design will convert the abstract vision of the application (in the ARB) into a concrete representation of the real deployment in the infrastructure.

During the process of creation of the final application structure, the platform can manage different options for the distribution of the Application Components and Products:

- All application components (ACs) in the same Virtual Machine
- Each AC on independent VM
- One AC replicated on several VMs (to support the customer's QoS requirements). This implies the inclusion of Load-Balancers (LBs) depending on the AC characteristics (for example stateless, stickiness, etc.)
- Different multi-tenancy options, in which the same Product Instance Container is used by several ACs of different applications (keeping some degree of isolation for performance and security reasons).

For the first implementation, only the simplest option will be implemented: one AC per VM. During the next months more complex deployment alternative will be covered. Besides this, the platform will also evolve supporting the NaaS concept, the ability to configure complex virtual network to better support multi-tiers applications.

During this first implementation the platform will also perform some basic integration of cloud services (like User Context, Publish/Subscribe or Storage as a Service). In the future, more complex integration will be performed including the access to network operators' services.

One of the important aspects of the 4CaaSt platform, the automatic adaptation and scalability will be implemented during the second year of the project, but the initial design is also available in the public deliverables of the project.

5. Conclusions and Future Work

This chapter provides a general overview of the 4CaaSt platform, the main motivations and objectives of the project, a general description of the 4CaaSt architecture and an overview of the current implementation status and future evolution.

During this first year, the project has focused on the design and some initial implementation of the platform, showing the most basic features for the 4CaaSt marketplace, platform management, cloud services, monitoring systems and the initial specification of the Application Blueprint language.

During the coming years, the project will continue working on a reference implementation of the 4CaaSt platform, focused on providing more advance features in terms of adaptation, scalability, multi-tenancy support, increasing the marketplace functionalities, and the integration with cloud products and services.

The project will also advance the component definition and specification of the 4CaaSt architecture that will be available on the project website (www.4caast.eu), together with the core components of the platform that will be publicly available under open-source license.

The cloud market is currently a very active area with big players pushing to move into the PaaS space (like Red Hat or VMWare), in this environment 4CaaSt will provide an innovative PaaS solution with unique features for PaaS Providers and Service Providers/Consumers in an open environment.

Acknowledgements

The authors would like to thank the following research projects: 4CaaSt (European Commissions IST priority of the 7th Framework Programme under contract number 258862), MEDIANET (Comunidad de Madrid S2009/TIC-1468), HPCcloud (MICINN TIN2009-07146) and FI-WARE (large scale integrated project funded under the EC FP7 ICT Work Program as part of its Future Internet Public Private Partnership initiative, http://www.fi-ware.eu, contract number 285248).

References

Assuancao et al. "Evaluating the cost-benefit of using cloud computing to extend the capacity of clusters." Procs HPDC2009, 2009.

Banga et al. "Resource Containers: A New Facility for Resource Management in Server Systems." Proceedings of the 3rd Symposium on Operating Systems Design and Implementation (OSDI'99), 1999.

Carzaniga, A., Rosenblum, D. S. and Wolf, A. L. "Design and Evaluation of a Wide-Area Event Notification Service." In *ACM Transactions on Computer Systems* 19 (3) (2001): 332–383.

Cugola, Gianpaolo and Munoz de Cote, J.E. 2005. "On Introducing Location Awareness in Publish-Subscribe Middleware." In Proceedings of the 25th International Conference on Distributed Computing Systems Workshops (Fourth International Workshop on Distributed Event-Based Systems), 6–10 June 2005, Columbus: IEEE Computer Society Press, 2005.

Czajkowski, Grzegorz and Von Eicken, Thorsten. 1998. "JRes: A Resource Accounting Interface for Java." Prodeedings of the 13th ACM SIGPLAN Conference on Object-Oriented Programming Systems, Languages and Applications, 1998.

Elmroth, Erik, Galán Márquez, Fermín, Henriksson, Daniel and Perales Ferrera, David. "Accounting and Billing for Federated Cloud Infrastructures." 1st International Workshop on Grid and Cloud Middleware (GCM2009). In proceedings of 8th International Conference on Grid and Cooperative Computing, GCC 2009a

Elmroth, Erik, Henriksson, Daniel, Galán Márquez, Fermín and Perales Ferrera, David. "Accounting and billing in the RESERVOIR federated cloud infrastructure." Internet of Services 2009, Brussels, June 10, 2009b.

Eugster, Patrick T., Felber, Pascal A., Guerraoui, Rachid and Kermarrec, Anne-Marie. "The many faces of publish/subscribe." *ACM Computing Surveys* 35 (2) (2003): 114–131.

Gilles, Floria, Hoyer, Volker, Janner, Till and Stanoevska-Slabeva, Katarina. "Lightweighted Composition of Ad-Hoc Enterprise-Class

Applications with Context-Aware Enterprise Mashups." 1st International Workshop on User Generated Services, 2009

Herzog, Almut and Shahmehri, Nahid. "Problems Running Untrusted Services as Java Threads." IFIP International Federation for Information Processing, 2005

Jacobsen, J. "The PADRES Distributed Publish/Subscribe System, Handbook of Research on Emerging Rule-Based Languages and Technologies: Open Solutions and Approaches." IGI Global, edited by Dragan Gasevic, Adrian Giurca, Kuldar Taveter, to be published 2010.

Jerzak, Zbigniew and Fetzer, Christof. "BFSiena: a communication substrate for StreamMine." In DEBS '08: Proceedings of the second international conference on Distributed event-based systems, pages 321–324, Rome, ACM, Italy, 2008.

JSR 154. "Java Specification Request 154: Java Servlet 2.5 Specification." http://jcp.org/en/jsr/detail?id=154 (accessed October 12, 2011).

JSR 220. "Java Specification Request 220: Enterprise Java Beans web site". http://jcp.org/en/jsr/detail?id=220 (accessed October 12, 2011).

Lenk, Alexander, Klems Markus, Nimis Jens, Tai Stefan, and Sandholm Thomas. "What's inside the Cloud? An architectural map of the Cloud landscape." In Proceedings of the 2009 ICSE Workshop on Software Engineering Challenges of Cloud Computing (CLOUD '09). IEEE Computer Society, Washington, DC, USA, 2009.

Leymann, Frank. "Cloud Computing". Proc. 52th Photogrammetric Week, 2009.

Mell P. and Grance T. "NIST Definition of Cloud Computing". Version 15, 2009, available online at http://csrc.nist.gov/groups/SNS/cloud-computing (accessed October 12, 2011).

Merrill Lynch. "The Cloud Wars: $100+ billion at stake." May 7, 2008.

O'Reilly Radar. "Web 2.0 and Cloud Computing." 2009.

OSGi. "OSGi Service Platform Core Specification". http://www.osgi.org/Release4/HomePage (accessed October 12, 2011).

Rochwerger, B., Breitgand, D., Levy, E., Galis, A., Nagin, K., Llorente, I., Montero, R., Wolfsthal, Y., Elmroth, E., Cáceres, J., Ben-Yehuda, M., Emmerich, W., and Galán, F. "The RESERVOIR Model and Architecture for Open Federated Cloud Computing." *IBM Journal of Research and Development* 53 (4) (2009): 4:1-4:11.

Stanoevska-Slabeva, Katarina and Wozniak, Thomas. "Cloud Computing Basics." In *Grid and Cloud Computing—A Business Perspective on Technology and Applications*, edited by Stanoevska-Slabeva, K., Wozniak, T., Santi, R. New York: Springer, 2009.

Thanos, George, Agiatzidou, Eleni, Courcoubetis, Costas and Stamoulis, George D. “Grid and Cloud Business Models.” In *Grid and Cloud Computing—A Business Perspective on Technology and Applications*, edited by Stanoevska-Slabeva, K., Wozniak, T., Santi, R. New York: Springer, 2009.

Vaquero, M., Rodero, L., Caceres, J. and Lindner, M. "A Break in the Clouds: Towards a Cloud Definition" *Luis ACM Sigcomm Computer Communication Review* 31 (1) (2009): 50–55.